Dark Tourism

Dark tourism, as well as other terms such as thanatourism and grief tourism, has been much discussed in the past two decades. This volume provides a comprehensive exploration of the subject from the point of view of both practice – how dark tourism is performed, what practical and physical considerations exist on site – and interpretation – how dark tourism is understood, including issues pertaining to ethics, community involvement and motivation. It showcases a wide range of examples, drawing on the expertise of academics with management and consultancy experience, as well as those from within the social sciences and humanities. Contributors discuss the historical development of dark tourism, including its earlier incarnations across Europe, but they also consider its future as a strand within academic discourse, as well as its role within tourism development. Case studies include holocaust sites in Germany, as well as analysis of the legacy of war in places such as the Channel Islands and Malta. Ethical and myriad marketing considerations are also discussed in relation to Ireland, Brazil, Rwanda, Romania, the UK and Nepal.

This book covers issues that are of interest to students and staff across a spectrum of disciplines, from management to the arts and humanities, including conservation and heritage, site management, marketing and community participation.

Glenn Hooper is a Lecturer in Tourism and Heritage at Glasgow Caledonian University, and has held academic appointments at St. Mary's University College Belfast, the University of Aberdeen and the Open University. He has published widely in travel and tourism, and is the co-founder of the international 'Borders & Crossings' Conference Series. His publications include *Land and Landscape, 1770–2000*, *Irish and Postcolonial Writing* (with Colin Graham) and *Travel Writing and Ireland, 1760–1860*.

John J. Lennon is the Vice Dean for the Glasgow School for Business and Society, Glasgow Caledonian University and Director of the Moffat Centre for Travel and Tourism Business Development. John has undertaken over 550 tourism and travel projects, in over 40 nations, on behalf of private sector and public sector clients. John is the co-author of *Dark Tourism: The Attraction of Death and Disaster* and a range of publications relating to the subject based on international research in the area.

New Directions in Tourism Analysis
Series Editor: Dimitri Ioannides, E-TOUR, Mid Sweden University, Sweden

Although tourism is becoming increasingly popular both as a taught subject and an area for empirical investigation, the theoretical underpinnings of many approaches have tended to be eclectic and somewhat underdeveloped. However, recent developments indicate that the field of tourism studies is beginning to develop in a more theoretically informed manner, but this has not yet been matched by current publications.

The aim of this series is to fill this gap with high quality monographs or edited collections that seek to develop tourism analysis at both theoretical and substantive levels using approaches which are broadly derived from allied social science disciplines such as Sociology, Social Anthropology, Human and Social Geography, and Cultural Studies. As tourism studies covers a wide range of activities and sub fields, certain areas such as Hospitality Management and Business, which are already well provided for, would be excluded. The series will therefore fill a gap in the current overall pattern of publication.

Suggested themes to be covered by the series, either singly or in combination, include – consumption; cultural change; development; gender; globalisation; political economy; social theory; sustainability.

A full list of titles in this series is available at: https://www.routledge.com/tourism/series/ASHSER1207. Recently published titles:

37 Being and Dwelling through Tourism
Catherine Palmer

38 Dark Tourism
Practice and Interpretation
Edited Glenn Hooper and John Lennon

39 Tourism Destination Evolution
Edited Patrick Brouder, Salvador Anton Clave, Alison Gill, Dimitri Ioannides

40 Advances in Social Media for Travel, Tourism and Hospitality
New Perspectives, Practice and Cases
Edited by Marianna Sigala and Ulrike Gretzel

Dark Tourism
Practice and interpretation

**Edited by Glenn Hooper and
John J. Lennon**

LONDON AND NEW YORK

First published 2017
by Routledge

2 Park Square, Milton Park, Abingdon, Oxfordshire OX14 4RN
52 Vanderbilt Avenue, New York, NY 10017

Routledge is an imprint of the Taylor & Francis Group, an informa business

First issued in paperback 2019

Copyright © 2017 Editorial matter and selection: Glenn Hooper and John J. Lennon; individual chapters: the contributors.

The right of Glenn Hooper and John J. Lennon to be identified as the authors of the editorial material, and of the authors for their individual chapters, has been asserted in accordance with sections 77 and 78 of the Copyright, Designs and Patents Act 1988.

All rights reserved. No part of this book may be reprinted or reproduced or utilised in any form or by any electronic, mechanical, or other means, now known or hereafter invented, including photocopying and recording, or in any information storage or retrieval system, without permission in writing from the publishers.

Notice:
Product or corporate names may be trademarks or registered trademarks, and are used only for identification and explanation without intent to infringe.

British Library Cataloguing in Publication Data
A catalogue record for this book is available from the British Library

Library of Congress Cataloging in Publication Data
Names: Hooper, Glenn, 1959- editor. | Lennon, J. John, editor.
Title: Dark tourism: practice and interpretation/edited by Glenn Hooper & John Lennon.
Description: Abingdon, Oxon; New York, NY: Routledge, 2016. | Includes bibliographical references and index.
Identifiers: LCCN 2016004649 | ISBN 9781472452436 (hbk) | ISBN 9781315575865 (ebk)
Subjects: LCSH: Dark tourism.
Classification: LCC G156.5.D37 D37 2016 | DDC 338.4/79104–dc23
LC record available at http://lccn.loc.gov/2016004649

ISBN: 978-1-4724-5243-6 (hbk)
ISBN: 978-0-367-36878-4 (pbk)

Typeset in Times New Roman
by Sunrise Setting Ltd, Brixham, UK

Contents

	List of illustrations	vii
	Notes on contributors	viii
	Introduction	1
	GLENN HOOPER	
1	**Is all tourism dark?**	12
	JOHN E. TUNBRIDGE AND GREGORY J. ASHWORTH	
2	**The long shadow: marketing Dachau**	26
	JOHN J. LENNON AND DOROTHEE WEBER	
3	**Prison tourism: exploring the spectacle of punishment in the UK**	40
	SARAH HODGKINSON AND DIANE URQUHART	
4	**Patrimony, engineered remembrance and ancestral vampires: appraising thanatouristic resources in Ireland and Sicily**	55
	TONY SEATON	
5	**Death camp tourism: interpretation and management**	69
	GREGORY J. ASHWORTH AND JOHN E. TUNBRIDGE	
6	**Guilty landscapes and the selective reconstruction of the past: Dedham Vale and the murder in the Red Barn**	83
	MARTIN SPAUL AND CHRIS WILBERT	
7	**A culturally constructed darkness: dark legacies and dark heritage in the Channel Islands**	96
	GILLY CARR	

8 **A light in dark places? Analysing the impact of dark tourism experiences on everyday life** 108
RIA DUNKLEY

9 **The undead and dark tourism: Dracula tourism in Romania** 121
DUNCAN LIGHT

10 **Genocide tourism in Rwanda: contesting the concept of the 'dark tourist'** 134
RICHARD SHARPLEY AND MONA FRIEDRICH

11 **Everyday darkness and catastrophic events: riding Nepal's buses through peace, war, and an earthquake** 147
SHARON J. HEPBURN

12 **From living memory to social history: commemoration and interpretation of a contemporary dark event** 160
ELSPETH FREW

13 **Experiencing dark heritage live** 174
BRITTA TIMM KNUDSEN

14 **Dark tourism in the brightest of cities: Rio de Janeiro and the favela tour** 187
GLENN HOOPER

Select bibliography 205
Index 215

Illustrations

Figures

2.1	Dachau Concentration Camp Memorial Site: grounds and signage, 2011	28
2.2	Overnight stay destination	31
3.1	The Galleries of Justice, Nottingham, incorporating Nottinghamshire County Jail	46
3.2	Death mask replicas of Brandreth and accomplices, Derby Gaol	50
4.1	'Natural history of two species of Irish vampires'. Ireland oppressed by two vampires – English landlordism, and the Church and Law	66
7.1	The links between dark events, legacies, heritage and tourism	98
7.2	The entrance posts of *Lager* Wick, Jersey	104
7.3	Information panel about *Lager* Wick, Jersey	105
9.1	Bran Castle in southern Transylvania (frequently mistaken for Castle Dracula)	124
10.1	Rwandan schoolchildren at the Kigali Genocide Memorial	139
12.1	Memorial Plaque and Visitor Centre, Lockerbie, Scotland	165
12.2	Quilt in Remembrance room and tree in Arts and Crafts room, Visitor Centre, Lockerbie, Scotland	167

Tables

2.1	Old Town and KZ tourist numbers	34
2.2	German and non-German tourist visitors	35
2.3	Dachau's marketing channels	36
6.1	The 'Managing a Masterpiece' Project: Project Outline (abstracted from Stour Valley Landscape Partnership, n.d.)	90
10.1	Visitors to Kigali Memorial Centre	135

Contributors

Gregory J. Ashworth was educated in geography at the Universities of Cambridge, Reading and London (PhD 1974), and has taught at the Universities of Wales, Portsmouth and, since 1979, Groningen, the Netherlands. Since 1994, he has been Professor of heritage management and urban tourism in the Department of Planning, Faculty of Spatial Sciences, at the University of Groningen. His main research interests focus on the interrelations between tourism, heritage and place marketing, largely in an urban context. He is author or editor of around 15 books, 100 book chapters and 200 journal articles. He received honorary life membership of the Hungarian Geographical Society in 1995 and an honorary doctorate from the University of Brighton in 2010, and was knighted for services to Dutch Science in 2011.

Gilly Carr is a Senior Lecturer in Archaeology at the University of Cambridge Institute of Continuing Education. She is also the Director of Studies in Archaeology and Anthropology and a Fellow of St Catharine's College. She has been a coordinator of the University's Heritage Research Group for the past five years. Dr Carr's research interests include POW studies and Conflict Archaeology, and she has published widely in these areas. Her fieldwork is currently based in the Channel Islands, where she has examined the complex and multiple legacies and victims of Nazi persecution during the German occupation of WWII. Recent published works include *Cultural Heritage and Prisoners of War: Creativity Behind Barbed Wire* (Routledge 2012), co-edited with Harold Mytum, and *Heritage and Memory of War: Responses from Small Islands*, co-edited with Keir Reeves. Her recent volumes include *Legacies of Occupation: Archaeology, Heritage and Memory in the Channel Islands* (Springer 2014) and *Protest, Defiance and Resistance: Channel Islands 1940–1945* (Bloomsbury Academic 2014).

Ria Dunkley is a Research Associate at the Sustainable Places Research Institute at Cardiff University, where her research focuses on Sustainable Communities. Ria's research centres specifically on building Sustainable Communities through community-led sustainability action in Wales and beyond. She has also held research roles at the University of Warwick and at the environmental charity, the Eden Project, in Cornwall. She completed her PhD in thanatourism, while her academic interests now include place-based sustainability initiatives, ecopedagogy, critical pedagogy and innovative qualitative research methodologies.

Elspeth Frew is an Associate Professor in Tourism Management and Discipline Head of Tourism, Hospitality and Events Management in La Trobe Business School, La Trobe University, Melbourne. Elspeth's research interest is in cultural tourism, with a particular focus on dark tourism and festival and event management. She recently co-edited two books, one on dark tourism and the other on the relationship between tourism and national identities. She has also conducted research into the relationship between the media and tourism management. Consequently, Elspeth's research is often inter-disciplinary, since she considers aspects of tourism within the frameworks of psychology, media studies, anthropology and sociology.

Mona Friedrich is currently a PhD candidate at the Institute for Dark Tourism Research (iDTR) at the University of Central Lancashire, exploring the complexities of memorialisation processes in Rwanda and the resulting implications of tourism development at sites of violence and death in post-conflict spaces. She previously completed a BA at the University of Sussex in Development Studies, and later completed an MA in Tourism, Environment and Development at the Geography Department of King's College London. She has published and presented her findings in the *Journal of Tourism and Cultural Change* as well as at various conferences, including the 2013 Peace Conference in Wageningen, Netherlands, and the 2015 Heritage of Death Conference in Stockholm, Sweden.

Sharon J. Hepburn is Associate Professor in Cultural Anthropology at Trent University, Canada. She has been conducting research on aspects of tourism in Nepal since 1990, most recently during the civil war between the Communist Party of Nepal (Maoist) and the Nepalese government security forces. Her tourism papers have appeared in the *Annals of Tourism Research, Fashion Theory, Food Culture and Society, Journeys*, and volumes edited by Michael Allen and Jonathan Skinner.

Sarah Hodgkinson is a Senior Lecturer in the Department of Criminology at the University of Leicester. She has a PhD in Psychology, specialising in the social psychology of aggression. Her current teaching and research interests include dark tourism, the psychology of 'evil' and forensic mental health. In 2013 she established a multi-disciplinary research network focusing on exploring Extremes of Human Cruelty. Her current research focuses on both Holocaust Tourism and Prison Tourism, within the context of the consumption of atrocity and tragedy, and the ethics of representation. She is a member of the Stanley Burton Centre for Holocaust and Genocide Studies, and has published in a range of international social science journals. She is currently co-editing *The Palgrave Handbook of Prison Tourism*, along with co-editors Jacqueline Z. Wilson, Justin Piché and Kevin Walby (2016).

Glenn Hooper is a Lecturer in Tourism and Heritage at Glasgow Caledonian University, and has held academic appointments at St. Mary's University College Belfast, the University of Aberdeen and the Open University. He has published widely in travel and tourism, and is the co-founder of the international 'Borders & Crossings' Conference Series. Glenn has also organised several international symposia relating to Travel Writing and the culture and history of Tourism and is the author of *Travel Writing and Ireland, 1760–1860* (Palgrave) and editor of *Landscape and Empire, 1770–2000* (Ashgate) and *The Tourist's Gaze: Traveller's to Ireland, 1800–2000* (Cork University Press). His work has appeared in the *Canadian Journal of Irish Studies, Literature & History, Eire-Ireland, The Journal of Design History, The Journal of Commonwealth Literature* and *Mosaic*, among others.

Britta Timm Knudsen is an Associate Professor in the Department of Aesthetics and Communication at the University of Aarhus. She has contributed to many international journals, including *Museum International, Journal of Tourism and Cultural Change* and the *International Journal of Heritage Studies*, and is co-author (with Carsten Stage) of *Global Media, Biopolitics and Affect: Politicizing Bodily Vulnerability* and co-editor (with Anne Marit Waade) of *Re-Investing Authenticity, Tourism, Place and Emotions*.

John J. Lennon is the Vice Dean for the Glasgow School for Business and Society, Glasgow Caledonian University and Director of the Moffat Centre for Travel and Tourism Business Development. John has undertaken more than 550 tourism and travel projects in more than forty nations on behalf of private- and public-sector clients. He undertakes research and commercial work in the fields of tourism development, destination marketing and financial feasibility of tourism projects. He is an Independent Specialist Policy Advisor to the Scottish National Tourism Organisation and VisitScotland, a Non-Executive Director of Historic Scotland and a former Board Member of the Canadian Tourism Commission European Marketing Group. John is the co-author of *Dark Tourism: The Attraction of Death and Disaster* and a range of publications relating to the subject based on international research in the area.

Duncan Light is Senior Lecturer in the Department of Tourism and Hospitality, Bournemouth University. He has a long-standing interest in Romania's ambivalent relationship with Dracula and its implications for tourism in the country. He is the author of *The Dracula Dilemma: Tourism, Identity and the State in Romania* (Ashgate 2012), is a contributor to the *Palgrave Handbook of Contemporary Heritage Research* (2015) and has published on Dracula tourism in *Annals of Tourism Research, Tourist Studies*, and *Journal of Dracula Studies*.

Tony Seaton is Emeritus Professor of Tourism Behaviour at the University of Bedfordshire and MacAnally Professor of Travel History and Tourism Behaviour at the University of Limerick, Ireland. He has taught and researched in the fields of travel and tourism behaviour and cultural studies at five British universities over thirty years, specialising in travel and tourism behaviour, travel history, Thanatourism (which he named in 1996) and literary tourism. His academic research and consultancy has been carried out for the Miller Library in Glasgow, the Beckford Society, the Natural History Society of Durham and Northumberland, and the MacAnally Travel Archive at the University of Limerick (for which he has written a bibliographical guidebook). His industry work has included consultancy and research in tourism for the United Nations World Tourism Organisation, the European Union, the European Travel Commission and twelve national, governmental organisations, including VisitBritain and VisitScotland, and many regional bodies. He has written/edited five books and published more than ninety articles, book chapters and papers. He is on the editorial board of two international tourism journals, including the *Journal of Heritage Tourism*. His continuing and current research interests lie in the history of Thanatourism and monastic travel, travel in eighteenth century verse, historical iconography and representation of travel in graphic satire.

Richard Sharpley is Professor of Tourism and Development at the University of Central Lancashire, Preston. He has previously held positions at a number of other institutions, including the University of Northumbria (Reader in Tourism) and the University of Lincoln, where he was Professor of Tourism and Head of the Department of Tourism and Recreation Management. He is co-editor of the journal *Tourism Planning & Development*, a resource editor for *Annals of Tourism Research* and a member of the editorial boards of a number of other tourism journals. His principal research interests are within the fields of tourism and development, island and rural tourism and the sociology of tourism, and his books include *Tourism and Development in the Developing World* (2008, with David Telfer); *Tourism, Tourists and Society, 4th Edition* (2008); *The Darker Side of Travel: The Theory and Practice of Dark Tourism* (2009, with Philip Stone); *Tourism, Development and Environment: Beyond Sustainability* (2009); and *Tourist Experience: Contemporary Perspectives* (2011, with Philip Stone). A second edited collection on tourist experiences, *The Contemporary Tourist Experience: Concepts & Consequences*, was published in 2012 while, drawing on his experience in tourism education, his book *The Study of Tourism: Past Trends and Future Directions* was published in 2011.

Martin Spaul has taught at Anglia Ruskin University for more than thirty years, lecturing in planning and sustainability. He holds an MA and PhD in Philosophy from the University of Cambridge and an MSc in Town Planning from Anglia Ruskin. Martin has been widely published on a number of subjects, including systems analysis, sustainability, planning, tourism and history, and is the author of several articles and chapters on tourism, heritage and ancient woodland. More recently he has published on generic skills for local communities in *Town Planning Review* (2011).

John E. Tunbridge is Emeritus Professor of Geography at Carleton University, Ottawa, Canada, recently Adjunct Professor at Curtin University, Perth, Australia and Visiting Professor at Brighton University, UK, with both of which he maintains active ties. He graduated from St. John's College, Cambridge, took his

PhD at Bristol University and was Junior Research Fellow at Sheffield University before moving to Canada in 1969. He has subsequently taught at the University of New England, Armidale, Australia; Portsmouth (now) University, UK; and the University of (now) KwaZulu-Natal, Pietermaritzburg, South Africa. He has written on heritage topics for forty years and has published extensively in heritage tourism and related areas, inter alia co-authoring *The Tourist-Historic City* (1990, 2000), *Dissonant Heritage* (1996), *The Geography of Heritage* (2000) and *Pluralising Pasts* (2007), variously with Gregory Ashworth and Brian Graham.

Diane Urquhart is a research student in the Department of Criminology, at the University of Leicester. She has a special interest in the historical, macabre and supernatural elements of punishment. Her PhD research is based on a contextualisation of this, through an examination of the contemporary dark tourist. Her interests lie more broadly within the realms of prison and penal tourism, and their construction and consumption by society.

Dorothee Weber is an Events Fundraising Officer at youth development charity Ocean Youth Trust Scotland. After graduating from Glasgow Caledonian University with an Honours Degree in Entertainment and Events Management, she spent several years working at the award winning Glasgow City Marketing Bureau in the convention and business tourism department. Her remit included research and sales activity in the UK association market to continuously promote Glasgow as a business tourism destination and to deliver conference business to the city. Dorothee has since moved into the third sector and is now responsible for looking after the charity's fundraising events, developing corporate partnerships and growing social media communication.

Chris Wilbert is a senior lecturer in Tourism and Geography at Anglia Ruskin University, and visiting professor in Geography and Tourism at the University of Bergamo. His research has focused on migrant work and tourism, environmental politics, human-animal geographies, and heritage tourism. He is co-editor with D. White of the book *Technonatures: Environments, Spaces and Places in the 21st Century* (Wilfred Laurier 2009) and *Autonomy Solidarity Possibility: The Colin Ward Reader* (AK Press 2011).

Introduction

Glenn Hooper

Emergence and growth

'Dark tourism', both as a category and as an analytical tool, has developed considerably in the past two decades. Lennon and Foley's *Dark Tourism* (1996), Tunbridge and Ashworth's *Dissonant Heritage* (1996), Ashworth and Hartmann's *Horror and Human Tragedy Revisited* (2005), Sharpley and Stone's *The Dark Side of Travel* (2007), as well as individual articles by Seaton and others reflected a growing interest in the field that stemmed in part from increasing access to sites in countries in eastern Europe and beyond.[1] Many more publications, sometimes developing discipline-specific lines, have since been added to the list, including Skinner's *Writing the Dark Side of Travel* (2012) and White and Frew's *Dark Tourism and Place Identity* (2013).[2] Not confined to dry and academic discussion only, the tourism industry, where interpretation and theory is tested and, if thought viable, absorbed and developed further, is no slouch when it comes to new ideas and opportunities. But an alliance between terms such as 'tourism' and 'dark'? What sort of recalibration was required to pull those two, seemingly incongruent, terms together? What did they mean; was it realistically possible to combine leisure with commemoration; and, in any case, what sorts of dangers might arise from a union of terms that ran so dangerously close to cancelling one another out?

Since many consider tourism itself to be a discipline of slippery and evasive classifications, responsive to the whims of international markets and consumers, it might be argued that the prefix 'dark' could provide a stabilising influence. As geographers, anthropologists and historians can attest, tourists can themselves be a varied and often complicated group. Out for good times, as often as not indifferent to all but their own private satisfactions and desires, they have often wreaked havoc with their demands and cultural insensitivities, their appetites and gluttonous capacity for superficiality and pleasure. Congested roads, litter and pollution, over-developed sites with their poorly constructed and indifferently designed hotels, exhausted infrastructure and bloated termini, not to mention the compromised conservation principles that surround many heritage sites (we will pass silently over 'visitor interpretation centres') – tourists are responsible for a great deal of cultural and environmental destruction. Yes, they bring in money, stimulate the local economy and are indirectly responsible for ensuring the implementation of new tourism and hospitality training programmes, but if ever there was a market of mixed blessings then this is it.

However, despite the mixed reviews and sometimes legitimate concerns expressed by tourism-dependent economies, whatever the criticism about the future of the industry and its insatiable demands, we know that tourism has brought good as well as ill. For example, in recent years, because of increasing green and environmental awareness, a genuine commitment to regeneration (rural and urban both) and sustainable programmes that work to a set of longer-term principles, concerned as much with tourism legacies as with immediate tourism impacts, we find genuine signs of hope. It is true that we are witnessing ever greater levels of market fragmentation, with tourism boards and operators proliferating wildly, but the fact remains that tourism can also be a positive. Recently developed tourism offerings in Namibia, for example, draw tourists not just to see wildlife and landscape, but also to become more aware of the demands for sustainable conservation programmes, including the need for animal welfare, new training initiatives and so forth.[3] Tourism volunteering programmes, such as those that are offered to visitors with interests in archaeology, are a well-developed part of the heritage tourism offer in Britain, much respected for the training offered, the experiences gained and the professional commitment of their staff.[4] The international Eco-Tourism Society offers rainforest cruises that bring hard currency to areas much in need of it, but in such a way as to impinge lightly or not at all upon the environment, with trips organised to the Galapagos, the Panama Canal and the Amazon, all sites high on the wish-list of adventure tourists. Combined with greater cultural understanding, much aided by an increasingly globalised and savvy constituency, the modern tourist is today a much more complex individual, part of an even longer and more diverse spectrum than was available thirty or forty years ago, but with the capacity to make a positive contribution nonetheless.

So when dark tourism was added to the mix, it might have been felt that here was a potentially useful development – that tourists who were interested in the memorialisation of the dead, who were concerned with historical atrocity and evil and driven by a desire for education and greater self-awareness, might serve the industry well. This cohort were reverentially attentive, less concerned about the trivialities of comfort and leisure than the average annual holidaymaker, and a relatively new category of international visitor. Gathered before war graves and around battle sites, at places of genocide or natural disaster fields, here was a different type of tourist, like nothing before seen, who was interested in weakness and failure and the human capacity for malevolence. Moreover, when Lennon and Foley produced their co-authored *Dark Tourism* they described in vivid detail the places where such tourists were now congregating in greater numbers, evidence of a movement towards increasing tourist sensitivity and concern: sites of trauma such as Changi Gaol in Singapore, associated with many atrocities perpetrated by the Japanese upon British and Allied soldiers; Auschwitz and Majdanek in Poland, the notorious death camps of the Third Reich; British memorials at the Somme; border conflict zones in Cyprus; American memorials at Pearl Harbor, Hawaii and others. Although Sharpley and Stone introduced greater thematic variety in their

The Darker Side of Travel, much of the emphasis still remained firmly associated with the historically accurate, where emotional and psychological depth was emphasised, and sites associated with trauma and barbarity highlighted. Terms such as 'kitschification' – a growing acceptance of the appeal of fear, anxiety and fun, of a clearly identified lighter side of dark tourism, in keeping with the growth of the term, the wider options available and the increasing interest being generated among tourists and providers both – were also foregrounded. However, much of the emphasis remained focussed on places of morbidity and grief, of torture, brutality and human suffering: the House of Terror in Budapest, the World Trade Center in New York, the Museum of Genocide at Tuol Sleng in Cambodia. Site managers might fret over the additional responsibilities that accompany dark tourism membership, but as for the dark tourists themselves, surely there could be nothing untoward about travellers whose interests centred mainly upon death or disaster?

In the debate about the emergence of dark tourism, and given the difficulties inherent in marketing human grief and suffering, it is important to remember that this niche offering in fact has a long history. Most commentators on dark tourism point to tourists from an earlier era, who were just as vulgar and tactless as any other travellers, and just as easily attracted to salacious thrills and excitement. Education or self-improvement were hardly evident in visits to see slaves fight to the death in the coliseum, to watch public executions and punishments in early modern England, to gawp at figures in asylums such as Bedlam or to look at sites of disaster where people drowned, were burned or fell to their deaths. It is difficult to interpret these impulses as more than the simple gratification of curiosity or, should we wish to put a more profound metaphysical gloss on it, for the purposes of considering their own mortality. Dark tourism has always existed in some form or other. What did not exist was the term itself, the marketing that has now grown up around it, the academic discourse such as we are presently engaged in and, most of all, the acceptance of the term into common usage. And while tourism boards and authorities might be cautious about a public engagement with the term, its connotations too much in conflict with the idea of spontaneous abandonment, relaxation and pleasure conjured up by the term 'tourism', there are still many for whom the term evokes mystery, excitement and the forbidden.

Ethics and reconsiderations

The opening up of trauma sites in eastern Europe and Asia has driven an academic and industry desire to identify more precisely what dark tourism actually encompasses. The field is required to respond to the ethical issues that arise in management and marketing and to engage in discussion about how dark sites might be simultaneously economically viable, and not overly exploitative of the trauma that the site contains. Building on work by Urry and Rojek in the 1990s, Lennon and Foley's *Dark Tourism* originally evaluated what was seen as a relatively new phenomenon: the desire by tourists for new experiences, especially experiences that related to death, disaster and atrocity. The authors provided examples of dark tourism sites as part of their analysis, but they also indicated that there were a range of other possibilities, including the 'former concentration camp, battle site, assassination or killing site or the location of a disaster', all of which could become 'a tourism resource to be exploited like any other'.[5] To help assess the recent upsurge of interest for such commodities they defined dark tourism as another example of late modernity, where everything is available for sale and consumption, including images and narratives associated with death. They also noted historic patterns of dark tourism that included visits made by tourists to 'cemeteries, mausoleums, churchyards', where death was commemorated and where issues of human mortality would therefore naturally arise.[6]

Almost a decade after the publication of Lennon and Foley's text there appeared another landmark publication in the development of dark tourism studies: Sharpley and Stone's *The Darker Side of Travel*, which brought together further analysis, a wider spectrum of case studies and, perhaps just as importantly, more questions about the entire dark tourist project, including an evaluation of the various ethical, management and interpretation issues that surround such sites. One of the most important issues that faced Sharpley and Stone centred on questions of definition, which Sharpley tackled rigorously in a perceptive introductory essay. There was increasing evidence of a rise in supply and demand, he argued, and the study of dark tourism was now described as 'both justifiable and important'.[7] However, the academic literature available was also 'eclectic and theoretically fragile', a number 'of fundamental questions with respect to dark tourism remain[ed] unanswered', and several earlier attempts at working in the area 'lacked theoretical foundations' and were 'largely descriptive' in content.[8] It would appear that although Lennon and Foley drew a line underneath what constituted dark tourism in 1996 (that it should be post-1900, preferably within living memory and capable of creating anxiety and doubt), the field had blossomed in the intervening years, to the point where it could include more diverse jurisdictions and examples and cross a much wider set of historical periods. Dark tourism, in other words, was now an academic growth area. Picked up by the media as well as academic publishers, it had also spread across schools of tourism management in the first instance, but enjoyed an increasing presence within the social sciences and the humanities. And what Sharpley clearly recognised, as he surveyed the proliferating spectacle that lay before him, was that this was both its strength and its weakness.

Perhaps a greater concern to several of the contributors to the Sharpley and Stone volume was a growing understanding of the ethical issues surrounding dark tourism, from both the supply and demand sides. While the image of tourists standing patiently in queues to visit Anne Frank's house, or the Jewish Museum in Berlin, testifies to a level of concern and decency, a determination to ensure that regrettable events are never forgotten or written out of the historical record, not all of those who attend such venues do so with the best of intentions. As with the killing fields in Cambodia or Angola, or national cemeteries that commemorate war dead, for some visitors there is, in addition to the draw of the historical story that is being told, an element of voyeurism, possibly even of adventure in being close to events that are reprehensible and obscene. And whatever management systems are put in place, however careful the marketing, despite a strong educational element in the form of new technologies and a clearly identifiable heritage component in place, there is always the possibility that some are attending for quite personal, arguably even perverse, reasons. In other words, efforts to engage as honestly and directly as possible with painful human memories so as to ensure a respectful visitor engagement with the site cannot always be guaranteed. And one must ask, in the context of tourism development, whether visitors should in fact be 'required' to acknowledge historic pain and suffering? It is to be hoped that visitors to dark tourist sites will be mindful of the suffering that took place there, but it is another thing to require them to visit for the 'right' reasons.

This is one of the key ethical conundrums thrown up by dark tourism. Visitor attractions have now developed that specifically cater for those who hanker after thrills, in the form of purpose-built sites dedicated to dark tourism themes, or attractions that have – in the interest of generating greater revenues – refurbished their offer in such a way as to take advantage of the recent interest in dark tourism as a niche area with development potential. Tourism operators at Chernobyl, for example, cater for a wide range of tastes and interests and offer anything from private one-day to week-long tours of the site and its facilities. Indeed, so successful has Chernobyl become as a dark tourism destination that several companies now work this particular location, many marketing their tours around provocative headlines designed to shock: 'communist propaganda', 'military installation', 'radiation reconnaissance routes'; these terms act as additional drivers in marketing the site. The success of the strategy is all the more remarkable given that recent reports suggest a relatively low mortality rate from the tragedy, most recently estimated at 4,000 deaths from radiation-related cancers, as opposed to the original Greenpeace estimation of 93,000. This raises the question of whether the people who organise such tours do so responsibly and with the full historical facts to hand. Do they offer value-free and objective interpretations? Do the tourists who sign up for such tours do so because of a desire to understand the dangers as well as the possibilities of nuclear technology, and out of respect for the dead and still suffering? Most importantly, are either visitors or marketers required to hold themselves to high moral standards, or do they merely regard these places as an afternoon's holiday diversion? Such ethical considerations have always posed some of the greatest challenges to the dark tourism project, and yet such issues also show no signs of becoming definitively settled any time soon.

Practice and interpretation

Although dark tourism has been a term much discussed in the past two decades, this volume specifically addresses the subject from the point of view of both practice (how dark tourism is managed and performed, what practical and physical considerations exist at site) and interpretation (how dark tourism is perceived and processed, what sorts of motivations or ethical considerations it elicits). Multi- and interdisciplinary in approach, as well as international in scope, theme and approach, this collection brings together experts in the field of dark tourism studies, with practical and operational expertise, but it also includes researchers from the fields of heritage, cultural geography, landscape studies and tourism history. Written with the wider cultural contexts and broader impacts of dark tourist venues firmly in mind, including the use of comparative examples where appropriate, the chapters employ case studies wherever possible, irrespective of disciplinary orientation. Several of the contributions are based upon questionnaires and interviews, or are at least partly produced on the basis of surveys, while others depend upon existing secondary literature and auto-ethnographic analysis. Contributors also consider dark tourism across the widest of spectrums, from the palest of sites through to the darkest, from fictional and filmic deaths to the horrors of historical atrocity as depicted at times of war and revolution, or as a result of postcolonial upheaval and trauma.[9]

Several of the essayists engage with visitor motivation and, while working across a variety of venues, produce not dissimilar conclusions concerning the underlying motivations behind tourist travel to alleged dark sites. In a detailed discussion of tourism in relation to Holocaust sites, Greg Ashworth and John Tunbridge warn against too easy a working definition of dark tourism, reminding the reader that many other overlapping or hybrid terms cover similar ground: disaster tourism, battlefield tourism, victim tourism, danger tourism and atrocity tourism. They also point to the sobering effects of commodification, even at Holocaust sites, of the 'Schindler experience', and of the need found among some tourists for the 'exciting frisson of proximity to horror'.[10] Tourism products can make for all sorts of difficulties, especially for the managers of such venues, many of whom are concerned with conservation, education and site-appropriate behaviour; these remain, not surprisingly, some of the greatest challenges facing anyone working in the sector today. Ria Dunkley's chapter also engages with visitor motivation, asking if such experiences make people more reflective and, more specifically, whether such visits can raise understanding and improve tolerance levels. Like Ashworth and Tunbridge, she identifies the modern need to explore the unusual and the 'urge to commemorate' that seems so prevalent today, but asks if exposure to such narratives creates positive behaviour and raises understanding, or rather intensifies and disseminates tension. After a consideration of several venues she concludes that, if thoughtfully and scrupulously managed, dark tourism sites can indeed produce benefits, particularly in developing a sense of connection between communities – a conclusion echoed by Richard Sharpley and Mona Friedrich in their analysis of genocide tourism in Rwanda. From a thought-provoking overview in which they suggest that dark tourism is too limiting, vague and, in its overlap with other niche tourism areas, a sometimes uncertain category, Sharpley and Friedrich highlight the positive and beneficial outcomes of the Kigali site. Proposed by the Rwandan authorities for UNESCO World Heritage Site status, such genocide museums, with their emphasis on education, greater understanding and learning, are to be much admired, especially in terms of the wider cross-community objectives and in their engagement with schools and other stakeholders. Sharpley and Friedrich prove that those who take the trouble to attend such sites do so out of commitment and a compelling need to understand, and they refute the suggestion of deviance or voyeuristic pleasure.

Commodification and the need for sites to remain vigilant in the face of potential exploitation has been discussed in earlier works. In their analysis of the notorious prison sites of Alcatraz and Robben Island, for example, Strange and Kempa point out that some 'heritage industry commentators, concerned generally about the inauthenticity of popularized "theme park" history, have denounced tourism as an inappropriate and even immoral vehicle for the presentation of human suffering'.[11] Peter Tarlow has reminded us that however careful managers might be, there will always be 'groups of people who see these locations as tourism draws and travel to these sites for both reasons of curiosity, nostalgia and pilgrimage'.[12] Graham Dann has discussed a catalogue of rather dubious products, from tours through 'war-torn hot spots' in the Balkans and Beirut to the Milan operator who was offering in the early 1990s '10-day tours into Lebanon and Dubrovnik for the equivalent of £15,000'. Massimo Beyerle, managing director of the company who masterminded such operations, 'claimed that he had a ready market for trips into war zones'.[13] Given the level of concern about what is being offered, it is hardly surprising that questions constantly arise over suitability, judgement, the ethics of specific venues and the shadowy and morally hazy worlds where enticement and danger persist. Dark tourism may have been around for a lot longer than imagined – during the opening skirmishes of the American Civil War, picnics were offered on the hills overlooking the battlefields – but it would seem that it is just as uncompromising and unfathomable as ever it was.

In addition to concerns over commodification, possibly one of the greatest challenges facing certain dark sites is when the proximity of life to death is most acute. For example, at those venues where death, disaster or the macabre are exclusively located, promotion may be relatively straightforward. But where these sites are located in the midst of living communities, further complications, largely of an ethical nature, immediately surface. At places such as Ireland's Famine Museum in Strokestown, Co. Roscommon, the Titanic Quarter in Belfast or the Workhouse Centre in Portumna, Co. Galway – all potential dark sites – information is contained within a single building or a cluster of buildings.[14] More importantly, each site is self-contained, a museumised venue containing narratives and exhibits, staffed by curators and guides who occupy the site through the course of a working day. But promotion becomes much more complicated at places like townships and camps, a theme developed in part by this author in a discussion of Rio de Janeiro's favela tours. This issue also surfaces in an analysis by Sarah Hodgkinson and Diane Urquhart of penal tourism in Britain – the stories of death, execution and punishment situated among still 'live' communities of inmates, and all the more traumatic for that. The authors agree that such tours of prisons are not without controversy. The potential commodification of something unseemly for public consumption, the morbid theme-park possibilities, not to mention the potential stereotypes produced of inmates and prison regimes are hardly negligible. While an educative and conservation ethos may underpin many carceral tours, intellectualising and stabilising both the purpose and the narrative, the authors suggest that dangers remain, and therefore a careful management policy is crucial.

Dark tourism within a 'live' environment is also the subject of both Britta Timm Knudsen's and Sharon Hepburn's chapters. Both chapters also employ online material, from blogs and discussion pages, to tease out the complexities of dark tourism sites that are themselves relatively fluid, and with sometimes as great an online than actual lived reality. Identified by Lennon and Foley twenty years ago as having a significant impact on the development of dark tourism, technology is still a major issue, and in Knudsen's discussion it becomes central to an understanding of the dark tourist experience. For Hepburn the impact of traditional as well as social media transmission of live engagement is also important, but mainly because in her analysis tourists become actors or participants in dark events that are themselves ongoing and in process. Like Knudsen, Hepburn is also interested in the immediacy of events, and of the effect upon tourists of life lived close to danger and conflict.

When dark events are converted into other forms altogether, or are conjured from the imagination, quite different challenges can materialise, as Duncan Light argues in his chapter on Dracula tourism. Horror and the macabre – but, more importantly, the spectacle or threat of cruelty – is a staple of gothic literature, and this is nowhere more vividly the case than in Bram Stoker's *Dracula* (1897). It is hardly surprising that such an extensive business has now grown up around the sale of all things 'gothic', with tourism and its events-related sister – what Emma McEvoy calls 'the scare attraction industry' – a central component.[15] Light assesses the impact of *Dracula* upon dark tourism principally in Romania, but finds the literary tourist industry of Whitby, North Yorkshire more forward-thinking and developed. Meanwhile Tony Seaton, in a wide-ranging chapter on thanatouristic patrimony, focuses on what he sees as the failure of the Irish tourist agencies to establish a greater claim to the Stoker and *Dracula* industries. Seaton regards Ireland as the best place for a developed consideration of gothic and thanatouristic tourism, forming, as it does, a tributary to the Anglo-Irish literary tradition, and including Charles Maturin's *Melmoth the Wanderer* (1820), Joseph Sheridan Le Fanu's *Carmilla* (1871) and Oscar Wilde's *The Picture of Dorian Gray* (1890). Yet Seaton also discovers that Whitby, with its Bram Stoker Film Festival and bi-annual Gothic Weekend festival, has forged ahead. In both chapters *Dracula* and its potential tourism impacts are situated at the lighter end of the dark tourism spectrum, where horror, imagination and spectacle profitably overlap.[16]

When we think of international dark tourism sites, we realise how instantly recognisable they have become by their name alone: Hiroshima, Chernobyl, Rwanda, Auschwitz. Associated with major trauma, loss of life, atrocity and violence, the names command instant recognition. The long list now includes Lockerbie, the small Scottish town almost exclusively associated with the downing of Pan Am flight 103 on 21 December 1983, and discussed by Elspeth Frew in terms of site management, healing and transcultural grief. In Frew's analysis the need for commemoration, including the incorporation of creativity as an element in the bereavement process, must be carefully handled so as to accommodate diverse experiences and memories. This problem is also addressed in John Lennon and Dorothee Weber's chapter on the work of the Dachau marketing bureau. Dachau, another town with a name forever engulfed in tragedy, faces a marketing challenge that seeks to draw visitors to the infamous work camp, while also promoting the attractions of the town and the surrounding region. Lockerbie is known to us for one thing, and one thing only; Dachau, on the other hand, is increasingly developing two. In Lennon and Weber's chapter emphasis is placed on the work of the Dachau marketing bureau, from where the authors draw at least some of their data about visitor motivation and site interaction, but from where one irrefutable question, despite local business efforts, also stubbornly arises: how can an architecturally important, historically and artistically resonant venue that also happens to have been a Nazi camp sell itself as something else?

This raises another key issue in dark tourism: not all sites that have the potential to exploit their dark histories wish to do so. Some places prefer amnesia to advertising, and would be much happier if the narratives of disquiet for which they are also acknowledged were silently sequestered forever. Polstead, a small town in rural Suffolk, is hardly a name that conjures up feelings of remorse or dread – a forgotten part of England, representative of Constable country, it is the sort of place Nikolaus Pevsner and John Betjeman both would have appreciated for its simplicity and charm, and a town that even today numbers its residents in the hundreds. In Martin Spaul and Chris Wilbert's chapter an early nineteenth-century murder, which was quickly turned into a successful tourist attraction, forms the basis of a discussion of how living communities deal with dark tourism. They describe how a brutal murder initially drew large numbers of visitors seeking vicarious thrills by travelling to the murder site and the place where the body was hidden, but how the locals swiftly tired of the notoriety that the event conferred on their village. The efforts to live down, forget or simply ignore this part of their dark history and heritage is an example of how problematic dark tourism can sometimes be for living communities. But in a digital age, at a time when dark tourism has an established (and arguably growing) reputation, can such guilty landscapes be ever truly forgotten? Such a question is also partly covered in the chapter by Gilly Carr, as well as in the opening essay by John Tunbridge and Greg Ashworth. In Carr's chapter the Channel Islands are discussed in relation to their role during the Second World War, but more particularly in terms of their subsequent handling of sites associated with incarceration, punishment and death. The tension that emerges out of such dark histories, she contends, has led to sites being ignored because of the difficult histories they tell and the threat to normality that they bring. Public memory, it would seem, can be just as selective as the individual sort, and for much the same sorts of reasons: because of the need for privacy, escape or secrecy.

Dark tourism has been with us in various formats for centuries, yet recent scholarly work on the subject, and the industry efforts made to define the field, indicate that it remains a rich area of enquiry. Questions continue to be raised concerning site management, and the broadening of the subject to include demand as well as supply analysis has focused attention on an area of tourist activity that shows no signs yet of fatigue. This collection adds to the work of previous researchers, but also develops ideas first outlined by Tunbridge and Ashworth in their *Dissonant Heritage* – a text not specifically engaged with the term 'dark tourism', but which nevertheless dealt with 'human unpleasantness' as it impinged upon heritage and the wider tourism offer. Indeed, many of the questions and uncertainties raised by Tunbridge and Ashworth still exist today. An acceptance that 'atrocity faces problems of definition that are far more intractable than the usual academic delimitation of a topic', a realisation that memorialisation has the potential to 'provoke glorification' (and therefore potentially deepen rather than heal wounds) and an acknowledgement that 'the management and marketing of the heritage of atrocity is the cardinal dilemma of dissonant heritage' are all issues that remain unresolved.[17] It is therefore appropriate that this introduction concludes with a reference to the chapter in this volume by John Tunbridge and Greg Ashworth who, in addition to bringing us full circle, focus on issues that continue to absorb, disturb and provoke. Drawing on a series of case studies, they consider the unseemly fascination we continue to have with atrocity and death, and more particularly the interpretive and management dangers that emerge when we shift our focus from 'site' to 'experience'. In an analysis of several holiday islands that includes Jersey, Malta and Bermuda, they pose questions not just of dark tourism but of tourism itself, uncoupling the term from its prefix so as to bring it more readily into the light, to demand of it more fundamental truths. And they ask: is not all tourism, in some way or other, in some place or other, dark?

Notes

1 See J. Lennon and M. Foley, *Dark Tourism: The Attraction of Death and Disaster* (London: Cengage, 2000); J. E. Tunbridge and G. Ashworth, *Dissonant Heritage: the Management of the Past as a Resource in Conflict* (Chichester: Wiley, 1996); G. Ashworth and R. Hartmann, eds., *Horror and Human Tragedy Revisited: The Management of Sites of Atrocities for Tourism* (New York: Cognizant, 2005); R. Sharpley and P. Stone, eds., *The Darker Side of Travel: The Theory and Practice of Dark Tourism* (Bristol: Channel View, 2009); T. Seaton, "Guided by the Dark: From Thanatopsis to Thanatourism," in *International Journal of Heritage Studies* 2, no. 4 (1996): 234–44.
2 J. Skinner, ed., *Writing the Dark Side of Travel* (London: Berghahn, 2012); L. White and E. Frew, eds., *Dark Tourism and Place Identity: Managing and Interpreting Dark Spaces* (London: Routledge, 2013).
3 Wildlife and conservation tourism has been professionally developed in Namibia for the past twenty years and more. For some earlier publications see C. Ashley and J. Barnes, *Wildlife Use for Economic Gain: The Potential for Wildlife to Contribute to Development in Namibia*, Research Discussion Paper no. 12 (Windhoek: Namibian Ministry of the Environment and Tourism, September 1996); J. Barnes, C. Schier, and G. Van Rooy, "Tourists' Willingness to Pay for Wildlife Viewing and Wildlife Conservation in Namibia," in *South African Journal of Wildlife Research* 29, no. 4 (1999): 101–11. For a sample of more recent work, see J. Barnes, "Community-based Tourism and Natural Resource Management in Namibia: Local and National Economic Impacts," in *Responsible Tourism: Critical Issues for Conservation and Development*, ed. A. Spenceley (London: Routledge, 2010), 343–57.
4 For a recent overview of some of the issues raised by volunteer tourism, see A. M. Benson, *Volunteer Tourism: Theoretical Frameworks and Practical Applications* (London: Routledge, 2011).
5 J. Lennon and M. Foley, *Dark Tourism*, pp. 9–10.
6 Ibid, p. 4.
7 R. Sharpley, "Shedding Light on Dark Tourism," in *The Darker Side of Travel*, ed. R. Sharpley and P. Stone, p. 7.

8 Ibid, pp. 6–11.
9 See W. F. S. Miles, "Auschwitz: Museum Interpretation and Darker Tourism," *Annals of Tourism Research* 29, no. 4 (2002): 1175–8.
10 On issues relating to commodification, see A. C. Wight, "Philosophical and Methodological Praxes in Dark Tourism: Controversy, Contention and the Evolving Paradigm," *Journal of Vacation Marketing* 12, no. 2 (2005): 119–29; C. Ryan, "The Buried Village, New Zealand – An Example of Dark Tourism?" *Asia Pacific Journal of Tourism Research* 11, no. 3 (2006): 211–26.
11 C. Strange and M. Kempa, "Shades of Dark Tourism: Alcatraz and Robben Island," *Annals of Tourism Research* 30, no. 2 (2003): 387.
12 P. Tarlow, "Dark Tourism: The Appealing 'Dark' Side of Tourism and More," in *Niche Tourism: Contemporary Issues, Trends and Cases*, ed. M. Novelli (London: Routledge, 2005), 57.
13 G. Dann, "Children of the Dark," in *Horror and Human Tragedy Revisited: The Management of Sites of Atrocities for Tourism*, ed. G. Ashworth and R. Hartmann (New York: Cognizant, 2005), 244.
14 See M. T. Simone-Charteris, S. W. Boyd and A. Burns, "The Contribution of Dark Tourism to Place Identity in Northern Ireland," in *Dark Tourism and Place Identity*, ed. L. White and E. Frew (Abingdon: Routledge, 2013), 60–78.
15 E. McEvoy, *Gothic Tourism* (London: Palgrave, 2016), 3.
16 For further discussion, particularly within the context of 'attraction-focused' tourism, see T. Blom, "Morbid Tourism – A Postmodern Market Niche with an Example from Althorp," *Norsk Geografisk Tidsskrift* 54 (2000): 29–36.
17 J. E. Tunbridge and G. Ashworth, *Dissonant Heritage*, pp. 95, 112, 129.

1 Is all tourism dark?

John E. Tunbridge and Gregory J. Ashworth

An opening perspective

In the summer of 2014 a cruise ship called, like many others, at a Norwegian port. Its passengers were earnestly advised to take every security precaution when going ashore, including avoidance of areas away from the main tourist sites; routine, perhaps, but in a place as generally inoffensive as Norway? Let's briefly consider some implications: the likely desirable expansion and diversification of the tourist city is inhibited; a subtle 'othering' of the local population is implied; and the wider security of the tourism enterprise is brought into question, another of the ubiquitous reminders of our post-9/11 world.[1] Dark tourism? Perhaps not; but a frisson of dark tourist experience on a continuum to dark tourism – perhaps, indeed.

This chapter reflects such personal experience over many decades as much as conventional tourism wisdom. We have been tourists ourselves for long enough to know that tourism is ultimately an individual and idiosyncratic affair. Thus we look a little askance at typologies of tourism, whether they refer to ecotourism, dark tourism or any other alleged variant of the beast. Our primary background in heritage tourism, which cannot usefully be separated from 'other' tourisms and which is very often very 'dark', increases our scepticism regarding the validity of attempts to force our experiences into predetermined categories of tourist.

Critiques of dark tourism have indeed already appeared. This chapter will review and elaborate their salient points and will seek to develop those arguments further, drawing upon that personal experience which we believe can more roundly illuminate our opening question. To this end, three perhaps surprising cases, differently illustrating an arguable ubiquity of dark tourism, will be discussed. We are not attempting to invalidate the present book, however: whatever our conclusions and those of other authors here, there currently remains a good deal to be debated around 'dark tourism', an influential concept over the past twenty years which is now ripe for extensive review.

Dark tourism: conception, usage, critique

The compelling notion of dark tourism was introduced by Foley and Lennon, who applied it to tourism sites and resources with lugubrious character or associations and sought to classify a very wide and eclectic range of cases according to type, such as deliberate massacres, assassinations and other crimes and accidental disasters of various kinds.[2] Their work was contemporary with Seaton's concept of 'thanatourism', that is, tourism motivated by associations with death.[3] It also paralleled Tunbridge and Ashworth's *Dissonant Heritage*, which was concerned with the inherently contentious quality of the attribute that motivates so much tourism (among other things).[4] Hartmann has perceptively identified all three works as the products of the post-Cold War intellectual environment which took advantage of unprecedented tourism and scholarly access to countless sombre sites in the formerly Soviet eastern Europe, post-apartheid southern Africa and elsewhere;[5] he is certainly correct as far as the present authors are concerned, for although the seeds of *Dissonant Heritage* were planted a decade earlier, a then unimagined opportunity was fortuitously presented by the 1990s world to develop its ideas from a much richer resource base.[6] All three works gave rise to further refinements by their authors and others, taking on somewhat modified lives in the process, and remain influential today, Lennon and Foley's ideas having been elaborated in *Dark Tourism*.[7]

Ashworth and Isaac have specifically reviewed the conceptual trajectory of dark tourism.[8] They note the growth in tourism activities and scholarship which occurred in the years since the concept was introduced, followed by the proliferation of other authors' dark tourism site claims and 'shades of grey' based on various criteria, producing unending overlapping taxonomies of diverse sites. Thus dark tourism encountered simultaneous proliferation and dilution, even before the darkness of sites was questioned by belated recognition of demand-side motivations and experiential responses. With respect to these, Ashworth and Isaac point out the arbitrary nature of site classifications and the centrality of motivation, experience and emotion to perceived 'darkness'. Emotional responses to tourism sites will vary on spectra from light to dark, weak to strong, and are of variable duration and ultimately immeasurable; and there are of course various kinds of dark response – among them horror, disgust, unseemly fascination – leading to the question of how we determine what is experienced, as one or another dark sentiment. Other authors might add the nebulous concept of 'affect' as the environmental conditioning in which emotional responses, dark or otherwise, are generated.

The key point, however, is that to shift darkness from sites to factors such as experience or emotion is to individualise it; a range of emotions will arise between both visitors and non-visitors, perhaps reactively between individuals. Moreover, it is to individualise it sequentially, in the context of successive encounters, for an individual may be desensitised away from darkness or experience a Damascene conversion to its recognition from one encounter to another at the same site. However emotion is felt, it may be reflected in behaviour, whether or not this follows some officially intended script, perhaps interactively between individuals and site agencies. However, dark tourism behavioural responses beyond the relatively straightforward, such as repeat visits or visits to similar sites, are ultimately not measurable and could lead us into the realm of clinical psychiatry, with uncertain implications for site management.

Following the above logic, any tourism experience might be seen as dark by some person at some time. As Ashworth and Isaac point out, there is moreover the question of degrees of darkness, its shade, intensity and duration, which will be central to the argument as extended in this chapter: who is to say that a generally mundane or inoffensive tourism experience is not at least minimally experienced as dark by someone, at some time, because of some association quite unknown to others? In light of these many caveats, Ashworth and Isaac conclude that we cannot eliminate the dark from tourism any more than from life itself, and if the stimulating (and now popularised) concept of dark tourism is to be intellectually sustained it requires rigorous restructuring – as this volume may hopefully achieve.

We will extend the argument further below by pointing to contexts where darkness, minimal or more, may be a deliberately sought-after gratification; where it may be an unexpected and more or less incidental realisation upon encounter with a tourism situation; where it may arise further along what is arguably the same continuum, as an unsought and quite unwanted negative tourism experience; or where it may be a condition left behind by a tourism oblivious or indifferent to its creation – thus tourism creates the darkness, rather than darkness, or its perception, creating or influencing the tourism.

Heritage tourism and dissonance

The preceding discussion brings us logically to heritage tourism, which is often difficult to separate from tourism as a whole. It has long been observed that tourists motivated primarily by outdoor pursuits, as in mountain or marine environments, will often turn to heritage attractions along the way or during inclement weather.[9] Indeed, we have surely all done so. More important, heritage is widely regarded as the dominant motivator of tourism in its own right. But this, of course, calls into question the nature of heritage, and in particular raises the question of which – or, more sensitively, whose – heritage so motivates.

This in turn brings us back to dissonance in heritage, first discussed by Tunbridge and Ashworth to identify the intrinsic discordance which exists, at least potentially, between all heritage and some or all of those who engage with it at some time or other. Much dissonance relates to the economic use of heritage in tourism, although much also concerns extraneous issues such as its significance with regard to socio-political identity. It is worth noting that the extensive discussion of heritage dissonance in subsequent years has sometimes misconstrued dissonant as dark or otherwise specifically problematical heritage, thus inadvertently linking it with dark tourism more closely than the original authors had intended or foreseen.

In parallel with Ashworth and Isaac's discussion of sites versus emotive responses, however, there is currently an unresolved question concerning the nature of heritage, that is, whether it constitutes 'the stuff' (tangible or otherwise) or 'the meaning of the stuff'.[10] If we take the former interpretation, which is commonplace in tourism, we may find that tourist destinations such as battlefields, prisons or sites of execution do indeed exude 'dark' heritage to any reasonable beholder, and thereby generate 'dark tourism' by commonly prevalent (if not by all) interpretations.

When we consider meanings, however, we may find that the darkness lies not so much in the eyes of the visiting beholder as in the resident's: the heritage resources may be relics of colonisation or of other former occupancy by visitors whose heritage they accordingly reflect, but they may be irrelevant or dissonant to residents.[11] Moreover, if we interpret heritage as meaning, new realms of potential tourism darkness may emerge the closer we look: there are colonial buildings in southern Africa which may appear irrelevant to the present majority of residents, but in fact hold a very real and dark heritage meaning as places of injustice, even execution, and thus touchstones of active hostility towards visitors who identify with them. This is not to dispute that postcolonial heritage meanings may simply be different, without active dissonance, as buildings and other resources have taken on new lives and are just perceived differently.[12]

Similarly, dissonances of heritage meaning between visitors and residents may (or may not) exist where former residents have been expelled and return as visitors to homelands repopulated by an alien society – the classic scenario between Germans and Poles, in particular, which became so visible in post-Cold War eastern Europe. The elimination of Central Europe's substantial Jewish population between 1933 and 1945 demonstrates this situation on a continental scale. One case is typical of many. The longstanding Jewish quarter of Krakow, Kazimierz had lost all its Jewish inhabitants by 1945. The reconstitution of the borders of Poland after 1945 created many refugees from the now Soviet-occupied Eastern territories, many of whom were re-housed in the former Jewish ghetto. The increasingly necessary renovation and redevelopment of the area was rendered next to impossible for the next forty years in part because of the issues and difficulties of reinserting a Jewish heritage presence into an area not only without Jews but now with a strong Christian presence. Starting in the late 1980s, the revitalisation plan for Kazimierz renovated and reopened the seven synagogues, opened Jewish restaurants and souvenir shops, reintroduced some Hebrew lettering, developed a Jewish heritage tourist trail and built a distinctive new 'Jewish Cultural Centre'. The area became – especially after Spielberg's 1993 film adaptation of Keneally's book *Schindler's List*, which was shot in Kazimierz – a major tourism attraction. Such tourism, much of which is a 'roots' tourism from North America, can be seen as doubly dark. For the tourist, the interpretation stresses not the many centuries of Jewish life in Kazimierz, but its short, dramatic, violent ending in 1940–3. However, the tourism also generates dark emotions among many current residents of the quarter and the wider country. Not only does the tourism ignore Christian Kazimierz, but there is a palpable sense of blame attaching to the non-Jewish Polish population for their inaction and even, some suspect, anti-semitism. To this sense of heritage exclusion and misappropriation could be added the uncertainties of the potential claims of former property owners or their descendants.[13] The re-creation of the Jewish quarter, in this as in many other European cities, especially after 1990, can be a dark experience for tourists and residents alike, although for different reasons.

Dissonant heritage scenarios giving rise to dark tourism potential could be more extensively cited, but this is hardly necessary. In keeping with Ashworth and

Isaac's points discussed above, such is the emotive individuality in interpretation of meanings that one can unequivocally conclude that, since all heritage is ultimately dissonant, all heritage tourism is dark tourism to some group, or to someone, at some time.

The travails of host societies – and tourists

We have argued above that the negativities of heritage tourism for host societies may be enough to 'darken' it, at least for some of the hosts, some of the time. For these, the darkest dimension may be the tourist's interest in destination attributes which are alien or even repugnant to some or all of those that live there. Take the case of the Japanese-built 'Burma Railway', popularised to tourists through the 1957 film *The Bridge on the River Kwai*, at Kanchanaburi in Thailand. This site and the neighbouring 'Hellfire Pass' attracts especially Australian tourists, as it was built in part by Australian prisoners of war, many of whom were mistreated and died here. This heritage tourism clearly evokes dark emotions (anger, grief, empathetic suffering and the like) but also, judging by the flags and national symbols left by tourists, other emotions, which are not necessarily dark (such as pride or gratitude). However, consider the experience of two other groups. First, the Japanese are the largest contingent of foreign tourists, but they are physically and emotionally separated from the heritage that attracts the Australians. Second, the Thai government and local people have an uncomfortable relationship to this heritage, as Thailand was at best a passive bystander and at worst an active collaborator in the Japanese occupation and resulting atrocities. The involvement of the local government and people in the heritage tourism of the 'Burma Railway' is minimal: most of the museums and heritage sites are managed and financed by Australian agencies, who have an uneasy relationship with the local authorities, or by the Commonwealth War Graves Commission.[14] Thus tourists from different countries and residents find the heritage to be 'dark', but in a different way and for different reasons.

However, there are numerous problems aside from ideological heritage values which could generate dark perceptions of tourism among hosts. In affluent host societies, tourists' intrusions – now with their iPad cameras – into private back gardens and into parking spaces, with general crowding of amenities, may generate hostility immune to persuasion that employment-generation and support of transport facilities make it all worthwhile; perception is all-powerful, after all, and could these hostilities not lead tourism to be perceived as dark? In poorer host societies its economic benefits may be well understood, but tourism's stimulation of crime, prostitution, unattainable local aspirations and an emigration exodus may seriously destabilise the host society, to say nothing of globalising influences contaminating local cultural traditions. The 'ugly American' of yesteryear may become simply the 'ugly tourism' of today, and deviant behaviour is even seen, at least by some, as dark tourism's research frontier.[15] Nevertheless, as our opening paragraph hinted, tourism can be inherently a dark experience for tourists themselves: the freedom and fulfilment promised by the travel industry is always

Is all tourism dark? 17

constrained and may be darkened by stresses of crowding, health, security and personal safety.

Jersey: dark tourism delayed

We have considered above numerous arguments for the ubiquity of dark tourism, of which the most compelling may be the individuality of 'demand-side' perceptions of tourism attractions. As we all should know but frequently forget, a message sent does not equate to a message received, in terms of its meaning. Nevertheless it is instructive to consider different jurisdictional examples of 'supply-side' tourism messages relating (or not) to dark tourism, for the tenor of messages sent is likely to influence the range of responses engendered among those who ostensibly receive them. Of course, even small jurisdictions are unlikely to be entirely consistent in the messages they project from different sources, even simultaneously and even from the same source, and the range of their market receptions may be accordingly still more diverse.

The three cases below are all 'holiday islands' dependent on their tourism economies and accordingly upon how their amenities are projected and received; all developed this role as beach resort destinations but have sought to diversify their tourism into heritage. All three are familiar to the present authors, who are interested in their heritage tourism and thereby in the likeliest source of tourism darkness. The first has belatedly capitalised on an island-wide dark tourism resource; the second has recently developed such with respect to part of its population; the third has yet to acknowledge unequivocally any dark elements in its tourism offer. How the different resulting messages of darkness (even if locally consistent) impact upon their different markets' reception is, we reiterate, another matter. Likewise, any shades of darkness that might be construed in tourist-resident interactions, individual tourism experiences of whatever kind or otherwise, are other matters again.

Any recent visitor to Jersey who reviews the television crime series *Bergerac* (1981–91) will go through many episodes in puzzlement as to whether this avaricious, venal, tax-evading mini-state could be the same place that is today increasingly marketing a specific dark tourism, which was typically hardly mentioned in that era thirty years closer to its historical resource. A *Bergerac* review certainly shows presently familiar scenes and, on a small island, cannot avoid distant glimpses of sinister landscape features; and eventually an episode does make subdued reference, for its plot value, to the Occupation.

The paradox of tourism reclamation of a historical resource predating an earlier period when it was largely disregarded for tourism is not difficult to understand, however, having significant parallels with Second World War tourism elsewhere in Europe. In Jersey there was no conquest or liberation battle and no post-Cold War renewal of access to heritage resources; but there was pain, guilt and conflict among the generation which shared the Nazi Occupation with continental Europe, inevitably suppressing reference to what in any case would not have been widely regarded as historically interesting in the immediate postwar decades.[16] Moreover, at that time the British proximity of Jersey's beaches, its use of sterling and its duty-free attractions – together perhaps with its touch of foreign exoticism, albeit within a familiar British context – counted for much more than they do today, when more distant warmer destinations are readily accessible and amenable. Thus the turn from 'blue' towards 'grey' (heritage) tourism as a relatively recent competitive stratagem was as logical here as in the other cases discussed below and in various other jurisdictions.[17]

The difficulty with the heritage of the 1940–45 Occupation, as in the rest of Europe occupied by the Germans, is that the local population resisted, collaborated or, most usually, did nothing, which after liberation left a sharply divided society. Consequently, as in much of western Europe, the postwar consensus was a deliberate heritage amnesia designed to protect social cohesion, often at the family level. The Occupation was a topic best officially and individually not remembered or commemorated. It was only after the passing of the generation that had experienced it that following generations sought answers to the questions about what had happened. A serious complication in the case of the Channel Islands, as the only part of Britain occupied by the Germans, was that it was formally demilitarised by Britain, the government remaining in place and instructed to cooperate with the German military authorities in order to safeguard as far as possible the well-being of the local population. This created a situation quite different from most of occupied continental Europe. A clear-cut historical narrative of heroes and villains is far more nuanced and muddied in the Channel Islands. There was a resistance in Jersey, albeit involving dominantly non-violent sabotage and passive civil disobedience. Some 2,600 residents of Jersey were arrested on some pretext and twenty-two died in German camps; one was shot for releasing a pigeon with a message for England (as now recorded in the Jersey War Tunnels museum). The problem, however, is not just that many of these resulted from local informants, but also that such resistance was not only an act of defiance against the occupying power; it was contrary to the instructions of the British government inside and outside the islands. Small wonder that there was a reluctance after 1945 to probe into what had happened during this uncomfortable episode in which local officials, including local police – on the instructions of the British government – implemented the Occupation, including ultimately the arrest and deportation of some Jews.[18]

If this is not potentially dark enough, there is additionally the direct role of the British government. The evacuation in June 1940, although completely predictable, was muddled, hurried and incomplete. The demilitarisation was not communicated to the Germans, who bombed the islands, leaving forty-four dead. The instructions given to Channel Island officials were ambiguous and unthought-out, and worst of all, in its consequences for the civilian population of the Channel Islands, was the decision not to liberate them by force after the invasion of the Continent in June 1944. Although no doubt justifiable on military grounds, given the formidable German fortifications, Churchill's decision to 'let them rot!' resulted in food shortages unresolved until the International Red Cross delivered emergency aid in December. The now prominent remembrance of the Red Cross ship *Vega* in museum, tapestry and postage-stamp expressions could be taken as a silent rebuke to Britain. There is a heritage here that is, to say the least, unhelpful in the at times fractious relationship between the Channel Islands and the UK.

Nevertheless, as Carr has documented, in Jersey's capital St Helier, the 'memorialscape' grew rapidly around the fiftieth anniversary of liberation, including belated recognition of victimised minorities, after decades of official forgetfulness.[19] Although the Channel Islands Occupation Society had been founded as early as 1961, the physical relics were largely treated as disposable scrap until the 1980s. The reclamation of a history now substantially beyond living memory as a valued heritage resource has been paralleled by the exploitation of the (virtually indestructible) German military installations; the imported labour, largely prisoners of war from the Eastern front, that built them; and their impact upon the occupied population, as a major tourism resource. The underground hospital has been progressively reincarnated (from an atypically early start) as the Jersey War Tunnels museum and other artefacts are now freely interpreted, notably the coastal 'Atlantic Wall' batteries and watchtowers which fit very well with active use of the coastal paths and cognate environmental interests. A central monument was unveiled in 1995 in what is now Liberation Square facing the St Helier harbour front in Jersey, and a series of other specific war-related memorials were located nearby in the 1990s and ensuing years.

Thus in Jersey we find the rehabilitation of formerly more marginalised historical relics as prime heritage resources for a now well-promoted, Occupation/military-focused dark tourism. The point is made here that 'supply-side' messages from official, and indeed voluntary NGO, sources can evolve through time to embrace – or, by implication, to withdraw from – dark tourism, whatever the 'demand side' makes of them, and whether or not it is indeed encouraged to form its own individual judgments of them. The overall 'demand-side' response is visually evident in the passage of tourists' feet, following the provision of information and means of access in which, ironically, the lead actor of *Bergerac* (John Nettles) plays a prominent interpretation role which draws upon his long attachment to Jersey in a rather different social climate thirty years ago. His BBC miniseries on the Occupation in 2010 and a subsequent book provoked a critical response in the islands, as it questioned the more comfortable clear-cut certainties of the Occupation. 'It is more morally complex, ambiguous and difficult. It is the story of a sustained and wholesale attack on human values, of great suffering, venality and violence.'[20] It could be dark indeed.

Bermuda: dark tourism polarised

Bermuda, a British overseas territory in the North Atlantic, has, like Jersey, belatedly discovered heritage tourism which has dark overtones. Unlike Jersey, however, it reflects a visible division within the population which aligns Bermuda closely with its giant neighbour and tourism market, the United States. Since we have discussed Bermuda elsewhere, we will do so more cursorily here.

Tourism in Bermuda was popularised in the late nineteenth century. Around 85 per cent of the visitors arriving by sea and 75 per cent of those arriving by air are from the United States, followed in importance by Canada and the UK. Its US dependence reflects both historic ties and, ironically, the British–US schism. Bermuda developed in close association with Virginia and evolved a North Atlantic trading economy. The wars between 1776 and 1814 destroyed this trade but substituted a defence economy. Bermuda became the British advanced base for military operations in the region, including the raids on Washington (1813) and New Orleans (1814). This strategic military role continued, albeit in a later British alliance with the United States, until well after the Second World War. The most important heritage resource resulting from this is the Royal Naval Dockyard, which could be considered dark for two reasons. Not only was its original purpose the conduct of war with the US, but it was constructed using forced, largely convict, labour.[21] The post-Cold War demise of the defence function has left Bermuda more dependent upon other economic activities which include selling the notion of 'paradise', based primarily upon the archipelago's subtropical marine 'blue' tourism imagery. It persists even though, as in Jersey, the competition of warmer and less expensive destinations (here the Caribbean) has eroded 'blue' tourism and prompted a turn to 'green' and 'grey', cultural heritage, tourism resources in an effort to sustain the industry.[22] Like Jersey, the main alternative economic resource is international finance, and with it the possibility of corporate tax evasion, itself dark in a different way.

However, 'grey' heritage resources have again evoked particular dark tourism shadows. Bermudian society is racially divided, like that of the American South with which it shares a common history of slavery, prior to the era of convict labour. Although emancipation came sooner to Bermuda, civil rights for the black 60 per cent of Bermudians did not; indeed, it was the colony's experience as a US wartime/Cold War base that exposed it to the civil rights agitation in the US and contributed to the ending of the 'colour bar' and establishment of universal suffrage. The legacy of race-based inequality dies hard and is reflected in the policies of opposing political parties. In this context the Progressive Labour Party, which recently held power, sought to promote the status of black Bermudians, *inter alia* by appropriately inscribing them into the historical record and therewith into heritage tourism. Thus 'paradise' became pockmarked by memorialisation of atrocities suffered during slavery and slave rebellions, if ameliorated by black achievements; there is now an African Diaspora Heritage Trail, a concept familiar to American visitors, linking sites across Bermuda which are largely unrelated to traditional tourism. And many of these sites would satisfy any definition of dark tourism.

A contrasting historical episode was Bermuda's use to accommodate around 5,000 Afrikaner prisoners of war during the 1899–1902 South African War, especially the 'bitterenders', the irreconcilable prisoners who were housed on Darrell's Island in the harbour of Hamilton, the capital. This has spawned distinctive narratives of resistance, escape and reprisals. Thus far the Royal Naval Dockyard appears not to be tarred with dark tourism, notwithstanding its construction initially by slaves and subsequently by convicts, and notwithstanding the recording of these histories on site by the National Museum of Bermuda. The Dockyard has been recast as a festival marketplace and prime tourism attraction, the appeal of which is mainly as a marine/leisure/retail environment geared to seaborne North American visitors. Its historical appeal is mainly as a mirror of US naval/military heritage, since it was constructed as a bastion against the young and hostile United States and served that role very effectively until the late nineteenth century and subsequent military cooperation. Some tourists will no doubt see this as dark; but the Dockyard surely contains the potential for a much darker, and locally divisive, heritage tourism if the suffering of those who built it is ever fully assessed.[23]

Malta: dark tourism deferred?

Malta received some 1.4 million tourists in 2012, largely from the UK, who stayed for 11.7 million nights with an average length of stay of eight days. This is dominantly a Mediterranean beach tourism. However, Malta's partial 'tourism turn' from its limited 'blue' to abundant 'grey' resources has engaged an almost unparalleled historical time-depth, from enigmatic Neolithic temples through the relics of Phoenician, Roman, Arab and Norman presence to the Knights of St. John, French and finally British occupations; all reflecting its pivotal location at the centre of the Mediterranean. A small island state lacks the means to nurture and develop this abundance of historical riches, and there has been a distinct general emphasis upon the Neolithic and Knights periods, commanding several World Heritage Sites. There is, however, a specific focus upon two momentous sieges – by the Turks against the Knights in 1565 and the Axis powers against the British/Maltese defence in 1940–3 – through both of which the defenders prevailed.[24]

There is currently little dark tourism in Malta from the supply side, however. The Neolithic remains are cast as 'mysterious' while the Knights are more personalised and celebrated for their splendid Baroque architecture set within massive fortifications, both epitomised by Valletta, which was built in the wake of the Turks' defeat. Both the sieges are typically presented in terms of heroic resistance – and with good reason – rather than as dark tourism experiences. The Axis siege would be particularly delicate to 'darken', since the villains of the time are, if not still living, in any case the forebears of the Italian and German tourists whose importance is second only to British visitors today. In addition, prewar Malta was subject to considerable fascist propaganda from its Italian neighbour, which did not entirely fall on deaf ears, and at the start of the conflict a number of local politicians and activists were arrested and deported, which does not enhance the 'George Cross' epic story of heroic resistance that is the dominant museum narrative. Of course, it is entirely possible for tourists and the Maltese themselves to receive the siege heritage narratives as dark, and there is little doubt that some do – particularly with respect to Malta's Second World War experience, which might well have dark more than heroic, personal or family associations, even to the point of motivating a tourist visit. There was personal suffering among the civilian population, with a substantial death toll, great privation in underground shelters and, by August 1942, starvation narrowly averted by the perceptually miraculous breakthrough of the 'Santa Marija' (Pedestal) convoy remnants. Equally, the entire British colonial heritage which culminated in that experience, though relatively little developed for tourism, could be received as dark by tourists of a fundamentalist anti-colonial persuasion, even though the Maltese do not present it as such, and it is worth remembering that Malta requested British protection in 1800 and in the era of decolonisation favourably considered incorporation into the UK (1956). A colonial struggle narrative is difficult to substantiate and, in the case of the Malta Heritage Trust (Fondazzjoni Wirt Artna) NGO, there is an active celebration of the British military history – as in Fort Rinella with its 100-ton Armstrong gun, and most visibly the noonday gun fired from Valletta's ramparts.

The 'Sovereign Military Hospitaller Order of the Knights of St. John of Jerusalem, Rhodes and Malta' (to give them their official title) represent a curious case of darkness deflected, or perhaps deferred, no doubt for the good reason of their marketability in terms of a romantic era which left much interest and beauty behind, along with their re-established diplomatic presence in Valletta and nearby Fort St Angelo. They offer a myriad of possible exciting and engaging narratives. We can leave aside the abundant conspiracy theory fantasies that credit the order (alongside the Templars) with almost every human misfortune, from the Iraq War to the assassination of JFK, as just rabid anti-Catholicism or even these days Islamic extremism.[25] There remains much to intrigue, excite and titillate, for these were no halcyon saints – even though they did defend Christian Europe from the then-perceived oriental infidel, the 'terrible Turk', whose status as villain is well established in the founding national mythologies of most of central and eastern Europe, through numerous encounters from the sixteenth to eighteenth centuries. The Knights were pirates and they were slavers, for a time involved in the West Indies sugar trade.[26] An impeccable Catholic source states: 'In this respect Malta remained a veritable slave-market until well into the eighteenth century.'[27] By the time Napoleon expelled them they had become corrupt and effete, so that the Maltese refused to take them back when, following Nelson's assistance in evicting the French, the British offered to reinstate them.

Thus Malta possesses dark heritage tourism potentials in some of its most prominently presented tourism resources, and would doubtless have no difficulty in finding dark narratives from its Roman, Arab or Norman as well as its Knights, French or British pasts, including its long naval role, which attracted enemy attention in war and privation through redundancy in peace.[28] An example of a muted dark narrative with greater tourism potential is recurrent slave-raiding by external pirates, which in the island of Gozo caused the inhabitants to flee to the central Citadel from time to time. Through much of Malta's history the role of the Maltese was at best that of supporting player and at worst passive victim of external powers, which weakens any contemporary nation-building.[29] Nevertheless, dark tourism may not sit well with a national heritage narrative that is overall positive and at times heroic. Even if it is not cultivated from the supply side, however, its presence in some tourists' 'gaze' is inescapable.

How dark, and for whom and what?

The above three cases reflect a particular kind of small destination and cannot simplistically speak for the global ebb and flow of tourism. However, they all reflect a much more widespread tendency in contemporary tourism: the quest to gain a competitive edge by diversifying their tourism from wider resource bases, capitalising on diverse, often 'niche', markets through a continuous diversification of the product line. This is particularly, but not exclusively, the case in long-established holiday destinations feeling the pressure of competition from warmer long-haul destinations now more readily and affordably accessible. Such diversification typically activates heritage attractions, within which niche opportunities may abound – and within which dark tourism attractions may well be offered or perceived. On a larger scale than the cases considered here, tourism moves to 'grey' may entail, or even equate with, moves – in various contexts and at differing paces – to 'dark', if only because so much of our common history lends itself to such interpretations.

We do not dispute that tourism may be a bright experience for many people much of the time. Our contention is, however, that it may constitute dark tourism by design or by circumstance, and it may do so for anyone at any time, whether or not by intent. Clearly there is a spectrum of tourism experience, from dreams through constraints to traumas, and with it a spectrum of contexts and encounters, ranging from the idyllic promise of the travel poster through the complexities of real experience to crossing a nebulous line (where – and whose?) into deepening shades of . . . dark tourism?

The cases we have discussed have all demonstrated dark tourism recognised over time, or potentially so, by most or some agencies, residents and/or visitors, essentially in the context of heritage dissonance. They have not demonstrated it in most of the other contexts discussed in this chapter simply because intrinsically they cannot: those other contexts are mostly too dependent upon individual experience and perception, and without extensive retrospective consumer surveys we cannot say whether dark tourism was anticipated or has arisen in respects such as stress and fear. But neither can we deny that it might have done so and might do so again for any particular group or individual, depending on the specifics of time, place and circumstance, if we are to press the concept of dark tourism to its ultimate conclusion. In so doing we note Hartmann's point, following Biran and Poria on deviant tourist behaviour, that the ultimate darkness of tourism may lie in its impact upon the global environment.[30] Since it has become the greatest migration in human history, leaving detrimental footprints from Mount Everest to the Poles, tourism – including ecotourism – merits further consideration from this darkness perspective than space permits here.

Dark tourism was recognised to be a broadening and diluting concept before this chapter was written. Extending earlier arguments, the intrinsic dissonance of heritage tourism enables us to assert that it may exist for any individual at any time in any place. This chapter has suggested that the darkness of tourism does not end even there. How far the further elements above are so implicated may be questioned; but this book is surely the place to frame the debate as to the ultimate reach of dark tourism.

Notes

1. G. J. Ashworth and J. E. Tunbridge, *The Tourist-Historic City* (Oxford: Elsevier, 2000).
2. M. Foley and J. J. Lennon, "Heart of Darkness," *International Journal of Heritage Studies* 2, no. 4 (1996): 195–7.
3. A. V. Seaton, "Guided by the Dark: From Thanatopsis to Thanatourism," *International Journal of Heritage Studies* 2, No. 4 (1996): 234–44.
4. J. E. Tunbridge and G. J. Ashworth, *Dissonant Heritage* (Chichester: Wiley, 1996).
5. R. Hartmann, "Dark Tourism, Thanatourism and Dissonance in Heritage Tourism Management: New Directions in Contemporary Tourism Research," *Journal of Heritage Tourism* 9, no. 2 (2013): 166–82.
6. J. E. Tunbridge, "Whose Heritage to Conserve? Cross-cultural Reflections upon Political Dominance and Urban Heritage Conservation," *Canadian Geographer* 28, no. 2 (1984): 171–80.
7. J. Lennon and M. Foley, *Dark Tourism* (New York: Continuum, 2000).
8. G. J. Ashworth and R. K. Isaac, "Have We Illuminated the Dark? Shifting Perspectives on 'Dark' Tourism," *Tourism Recreation Research* 40, no. 3 (2015): 316–25.
9. Ashworth and Tunbridge, *The Tourist-Historic City*.
10. J. Warren-Findley, "Rethinking Heritage Theory and Practice: The US Experience," *International Journal of Heritage Studies* 19, no. 4 (2013): 380–3.
11. Tunbridge and Ashworth, *Dissonant Heritage*.
12. E. van Maanen, *Colonial Heritage and Ethnic Pluralism: Its Socio-psychological Meaning in a Multiethnic Community: The case of Paramaribo, Surinam* (Breda: NRIT Media, 2011).
13. G. J. Ashworth, "Holocaust Tourism and Jewish Culture: The Lessons of Krakow-Kazimierz," in *Tourism and Cultural Change*, ed. M. Robinson, M. Evans, and N. Callaghan (Newcastle: University of Northumbria, 1996), 363–7.
14. G. J. Ashworth, "Ethnic Conflict: Is Heritage Tourism Part of the Solution or Part of the Problem?" in *Transformational Tourism: Host Perspectives*, ed. Y. Reisinger (Wallingford: CABI, 2014), 167–80.
15. A. Biran and Y. Poria, "Reconceptualizing Dark Tourism," in *The Contemporary Tourist Experience: Concepts and Consequences*, ed. R. Sharpley and P. Stone (London: Routledge, 2012), 59–70.
16. C. G. Cruickshank, *The German Occupation of the Channel Islands* (Guernsey: The Guernsey Press, 1975).
17. G. J. Ashworth and J. E. Tunbridge, "Changing Tourism Destinations from Blue to Grey Tourism: The Malta Example," *Tourism Recreation Research* 30, no. 1 (2005a): 45–54; 'Move Out of the Sun and Into the Past: The Blue-Grey Transition and its Implications for Tourism Infrastructure in Malta," *Journal of Hospitality and Tourism* 3, no. 1 (2005b).
18. M. Bunting, *The Model Occupation: The Channel Islands under German Rule, 1940–1945* (London: Harper-Collins, 1995).
19. G. Carr, "Examining the Memorialscape of Occupation and Liberation: A Case Study from the Channel Islands," *International Journal of Heritage Studies* 18, no. 2 (2012): 174–93.
20. J. Nettles, *Jewels and Jackboots: Hitler's British Channel Islands* (Jersey: Channel Island Publishing and Jersey War Tunnels, 2012). Quoted by J. Carpenter in 'Telling the Truth about Channel Islands Cost Me My Friends," *Daily Express*, November 5, 2012.
21. J. E. Tunbridge, "Large Heritage Waterfronts on Small Tourist Islands: The Case of the Royal Naval Dockyard, Bermuda," *International Journal of Heritage Studies* 8, no. 1 (2002): 41–51.
22. Ashworth and Tunbridge, "Changing Tourism Destinations," 'Move Out of the Sun."
23. J. E. Tunbridge, "Large Heritage Waterfronts'.

24 Ashworth and Tunbridge, "Changing Tourism Destinations," "Move Out of the Sun"; J. E. Tunbridge, "Malta: Reclaiming the Naval Heritage?" *International Journal of Heritage Studies*, 14, no. 5 (2008): 449–66.
25 E. Lebedev, "Caped Crusaders: What Really Goes On at the Knights of Malta's Secretive Headquarters?" *The Independent*, March 29, 2014.
26 J. Riley-Smith, *Hospitallers: The History of the Order of St. John* (London: Hambledon, 1999).
27 C. Moeller, "Hospitallers of St.John of Jerusalem', *The Catholic Encyclopedia* (New York: Robert Appleton Company, 1910), www.newadvent.org/cathen/07477a.htm
28 Tunbridge, "Malta: Reclaiming the Naval Heritage?".
29 J. Baldacchino, "Pangs of Nascent Nationalism from the Nationless State: Euro Coins and Undocumented Migrants in Malta since 2004," *Nations and Nationalism* 15, no. 1 (2009): 148–65.
30 Hartmann, "Dark Tourism'; Biran and Poria, "Reconceptualizing Dark Tourism."

2 The long shadow
Marketing Dachau

John J. Lennon and Dorothee Weber

Introduction

The town of Dachau lives in the shadow of its terrible past, known for its association with the Nazi regime and as the place of the former concentration camp. Although town officials have attempted to market the town's cultural and historic features, visitor numbers for Dachau's other sites have been a fraction of those of the memorial site/concentration camp. Using a quantitative intercept survey, the aim of this chapter is to investigate the reasons why tourists visit Dachau and consider what such findings mean for destinations with 'dark' history and pejorative place names. The study utilises a mixed-methods approach, triangulating a literature review, a pilot intercept survey and a number of key stakeholder interviews. The findings explore the relationship between Dachau's various identities as a town, a Holocaust site and tourist destination, and the impossible legacy it carries. Throughout the chapter, the term 'dark tourism' will be used as a general term to refer to the visitation of sites associated with recent death, disaster and atrocities, and KZ (the German abbreviation for 'Konzentrationslager') will be used to denote 'concentration camp'.

Holocaust site, tourist destination

The German town of Dachau is infamously associated with the atrocities committed during World War II (1939–45). Opened on 22 March 1933 on the deserted grounds of an unused gunpowder and munitions factory from World War I, Dachau was erected as the 'first state concentration camp', and served as a model and training facility for all other concentration camps.[1] Over the span of its existence, 200,000 people from all over Europe were imprisoned and more than 41,500 were murdered.[2] On 29 April 1945 the survivors of the camp were liberated by American troops, and during the liberation photographic and film images of the camp were taken and broadcast around the world. The distribution of these dreadfully iconic images meant that the name of Dachau became associated with some of the most extreme examples of Nazi persecution. Today, Dachau concentration camp is one of the most visited concentration camp memorial sites, attracting more than 800,000 visitors every year, although this number is a conservative estimate as there was no monitoring system in place at the site when research was undertaken in late 2011.[3] The estimate is based on two surveys of visitors to the site that shows that 618,000 audio guides were hired, although the number of visitors is likely to be considerably higher.[4] This chapter will examine visitor motivation in relation to this particular site, and situate it within the wider context of dark tourism. It will also carefully explore the difficulties marketing professionals face when attempting to show that there is more to Dachau than 'just' the concentration camp.

Dark tourism has its origins in the work of Lennon and Foley and can therefore be placed within the wider framework of tourism studies. However, literature directly referring to 'dark tourism' remains limited and predominantly descriptive in nature.[5] Lennon and Foley's study considered the ethical issues relating to the display and interpretation of sites and objects, and the nature of political and ideological messages communicated by interpretations at sites of dark tourism.[6] Earlier, Rojek had addressed the notion of such tourist attractions and introduced the concept of 'Black Spots',[7] which referred to the 'commercial development of sites in which [. . .] people have met with sudden and violent death'.[8] Following on from Rojek, Seaton described death-related tourist-activity as 'thanatourism', which he defined as the 'travel to a location . . . motivated by the desire for actual or symbolic encounters with death'.[9]

Of course, dark tourism is not a new phenomenon and sites of public executions, battles and natural disasters have long exerted their appeal. More recent manifestations of dark tourism range from war museums to war memorials and genocide sites,[10] often incorporating ideological elements in site interpretation and terms of remembrance.[11] Rojek argues that death-related sites can also be referred to as 'sensation sites', and that they reflect some aspects of what contributors have referred to as part of the 'society of the spectacle'.[12] In the case of Dachau, this sensation has been experienced through photographic and filmic imagery transmitted at the time of liberation and subsequently viewed by millions of people since. In this context the term 'dissonant heritage' is a useful way of describing the seeming incongruence between people's current lives and their heritage.[13] When atrocity has occurred in the recent past, this can manifest in a particularly intense manner as people struggle to make sense of, and interpret, what has occurred. When the location of the atrocity becomes an educational and/ or tourist site, the complexity of, and demand for, 'interpretation' increases.[14] Clearly, then, there are issues with the interpretation of such sites in terms of how they are constructed and managed, what is displayed and any accompanying political, social or philosophical messages utilised (see Figure 2.1).

Wollaston helpfully defines four roles of Holocaust museums: sites of mass tourism, memorials to the dead, vehicles of historical exposition and living classrooms. However, this can lead to conflicting contexts within the same site, for 'what is appropriate from an educational point of view may be considered inappropriate, even offensive, in a commemorative context'.[15] However, of possibly greater concern in this context is the issue of commercialisation, which has been commented upon with regard to numerous sites. Beech, for example, notes that the Buchenwald camp has a restaurant and a bookshop, and asks whether a clear

Figure 2.1 Dachau Concentration Camp Memorial Site: grounds and signage, 2011.
Photo copyright D. Weber.

line can be drawn between essential visitor services and what might be perceived as the unacceptable exploitation of other people's horror and tragedy.[16] Stone suggests that the interpretative messages that are conveyed to visitors can be further blurred by the commercialisation of the sites.[17] The relationship between the town of Dachau and the KZ, and the context of commercialisation, are both of particular interest here. In the past, Lennon and Foley referred to Dachau's 'intractable public relations problem', and described a sign at the entrance to the KZ inviting visitors to 'Visit Dachau, the 1200 year old artists' centre with its castle and surrounding park'.[18] Over the sign, 'HAVE YOU NO SHAME' had been written in English and German. In the realm of 'sightseeing in the mansion of the dead', issues of appropriateness become extremely difficult.[19] Interestingly, this has been raised in contemporary theatre and film. For example: in Alan Bennett's play *The History Boys*, a teacher talks about school trips to Dachau and asks, 'Where do they eat their sandwiches? . . . Do they take pictures of each other there? Are they smiling? Do they hold hands? Nothing is appropriate!' This, essentially, is the problem faced by the town of Dachau. In the shadow of its KZ, can it and should it market its castle, artists' centre and beautiful parks? Is anything appropriate in terms of destination marketing?

Located in Upper Bavaria in South Germany, the district town of Dachau is 18 km from central Munich (a 15-minute train journey) and has a population of about 40,000 inhabitants.[20] The Old Town of Dachau currently hosts an array of cultural offerings, including concerts held in the baroque castle, as well as exhibitions in its art galleries and museum. These features are used in marketing and promotion. However, despite the obvious attractions, for many the defining draw of the destination remains Dachau's KZ. As a site it has undergone numerous changes of use and political significance since its creation, and the establishment of a memorial site was far from predetermined. Moreover, the memorial site has also encountered difficulties in terms of 'authentic' representation and interpretation, while tensions have developed between its memorial and educational purposes.

Opened in 1933, Dachau developed to serve as a model for all other Nazi concentration camps, as a 'school of violence' where KZ commandants and officials received training[21] as well as a camp to accommodate political prisoners, including heads of state from occupied countries.[22] Ironically, German and foreign officials were frequently taken on tours of the camp and the areas viewed underwent elaborate preparations prior to such visits, to ensure that a 'clean' image of the camp was portrayed to the outside world. During the final years of the war (1943–5), overcrowding in the Dachau camp and the presence of numerous diseases made such tours increasingly difficult.[23] With the liberation of Dachau and its surviving prisoners on 29 April 1945, the shocking images and film footage of the reality of the KZ soon circulated worldwide.[24] From July 1945 to summer 1948, the occupying US Army used the former KZ facilities as an internment camp for arrested National Socialist functionaries and members of the SS. During this time and parallel to the Nuremberg trials, a US military court was set up in the grounds of the camp. Survivors were encouraged to install a temporary exhibition in the gas chamber crematorium, documenting the brutality of everyday life in the KZ. During the period of the Cold War that followed, the US began to 'relax its punitive stance toward Germany to gain favor of the German populace'.[25] The strategic importance of Germany's location became vitally important in the post-war struggle, and as a consequence the Dachau trials ended and surviving prisoners were repatriated to their home countries within two months of liberation.

With the arrival of thousands of German refugees from eastern Europe, Dachau KZ was converted into a residential settlement for refugees in 1948. The then Bavarian government did not see any necessity for a memorial, nor any educational purpose in preserving the KZ site, and many of the original facilities and artefacts were removed or converted to cater for the needs of the residents. Distel notes: 'in these years, there was essentially no interest in the history of the years between 1933 and 1945 and the fate of the victims of the National Socialist dictatorship in the vicinity of Dachau'.[26] On several occasions, Dachau's representative in the Bavarian State Parliament demanded the destruction of the crematorium to 'put an end to the "defamation of the Dachau area"'.[27] However, camp survivors stringently resisted these demands and joined together to form the Comité Internationale de Dachau. In the late 1950s they began a decade-long campaign which eventually led to the closure of the refugee settlement and the reconfiguration of the former KZ as a memorial site.[28] However, by 1964 the old camp barracks had been demolished and two barracks had to be rebuilt to represent the original buildings. As the memorial site was developed, many decisions were made regarding the reconstruction and representation of the former camp, illustrating the tensions such sites face when required to combine the functions of memorial/commemoration of victims, as well as documentation of the circumstances of their incarceration and treatment. Such reconstruction, clinical recreation of interpretive barracks, landscaping, and so on will invariably invite comment in relation to the degree of authenticity in respect of site buildings and fittings.

Between 1968 and the late 1990s there were few fundamental changes to the site and exhibition content. However, during this period a younger generation began to visit the memorial site. In turn a number of local groups, initially opposed by the authorities, began campaigning for accommodation and resources to encourage the education of such youth groups. This mounting pressure led to the eventual development of an educational resource at the KZ memorial site in the late 1990s. A group comprising historians, Dachau survivors and education groups planned significant renovations;[29] an 'advisory council' was organised which recommended that 'the planned visitor tour should retrace the path that the entering inmates followed, starting at the entrance gate and continuing into the service building where the initial registration took place, then proceeding to the barracks area and crematorium'.[30]

It is interesting to note that this prisoner journey remains the guiding theme for the memorial site and its exhibitions today.[31] Dachau is the second most visited KZ memorial site after Auschwitz-Birkenau, and, as noted above, is visited by an estimated 800,000 people per annum.[32] Indeed, there appears to be continued interest in the Nazi past among new generations, not least because of education and school programmes.[33] According to the most recent data the majority of visitors are international, although it should be noted that establishing exact visitor statistics has proven to be difficult due to access and the layout of the site.[34]

The future of the Dachau KZ memorial site is hard to determine as it moves to its next stage. The numbers of those with memories and lived experience of the Nazi era are steadily dwindling, although some have argued that this demographic evolution can be seen as positive, helping to develop places of learning that serve to create meaning in a contemporary context. Indeed, the complexities and tensions related to the memorial site cannot be considered in isolation from the town of Dachau, which provides an interesting focus for a further investigation. First, the town's identity is entwined with the concentration camp and the atrocities that took place there and, unlike other such camps, both the town and the Holocaust site carry the same name.[35] Second, prior to World War II, the town of Dachau was a centre for the arts, and painters travelled from all over Europe to be part of Dachau's artist colony.[36] Currently, the town of Dachau has a number of art galleries, exhibition spaces and museums, which are home to an array of cultural events throughout the year. Third, Dachau's identity is also impacted by its close proximity to the major tourist destination of Munich. These influences have historically put the tourist authorities of Dachau in an almost impossible marketing and destination development dilemma. Indeed, one might argue that any marketable identity for Dachau is invariably compromised, given its primary identity as a Holocaust site.

Since the 1990s, Dachau town's tourist authorities have attempted to market Dachau as offering something distinct from the KZ. A 2006 Dachau tourist promotion film challenges its viewers with the headline 'Wieviel Dachau kennen Sie?'

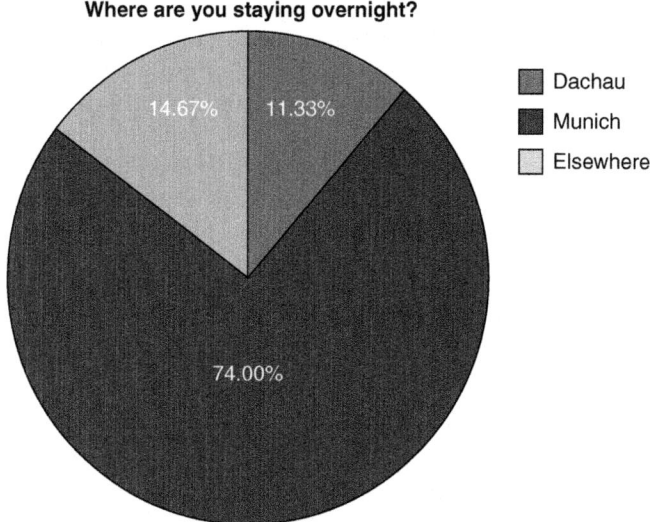

Figure 2.2 Overnight stay destination.

('How much do you really know about Dachau?').[37] In tourist brochures and on the town's official website (see Figure 2.2) great emphasis is given to Dachau as a town that has much to offer beyond the KZ, although there have been varied responses to these marketing efforts. A 2007 study, for example, found that only 8 per cent of visitors to the Holocaust memorial site could reference other associations they had to the name of the town.[38] Despite attempts to attract visitors from the memorial site into the Old Town by introducing a town centre stop on the shuttle bus route to the KZ, town officials found that very few people visited the Old Town and the service was discontinued in 2010.

As in other small towns, Dachau town officials were responding to the growth in urban tourism and accordingly developed a brand and tourism marketing strategy. However, in Dachau's case, the unique features that constitute the town's brand are based on its disparate historical record: from artist colony, to Holocaust site, to centre for education and international youth encounters. While many towns and cities have exploited the opportunities provided by tourism for economic gain, Dachau's marketing efforts have been focused on re-shaping the town's image.[39] As Baker has noted, 'a brand image relates to how the brand is perceived from the customer's point of view, while brand identity is the unique set of stimuli that express the brand and shape its image'.[40] However, for most potential visitors to Dachau, the best known representation is the synonymous KZ and the darkly iconic film and photo imagery of the liberation of the concentration camp. Kolb notes that in order to change a negative image, cities must disassociate themselves from their negative pasts, creating a new identity.[41] Former industrial cities such as Glasgow or Manchester were associated with high levels of deprivation and crime, but have undergone highly effective place re-branding, and repositioned themselves as cultural centres and urban tourist attractions.[42] In Dachau, tourist officials and city authorities continually struggle to disassociate themselves from the atrocities that took place for fear of accusations of revisionism, or diverting attention from the reality of the tragic past of this place. As a compromise, it would appear that tourist officials and civic authorities have presented the 'educational opportunity' as a unique positive selling point. However, the tension of cultivating the town's 'other' identity without erasing its associations with the KZ remains.

Understanding visitor motivation

The methodology applied in this chapter is a mixed-methods approach that utilises qualitative and quantitative elements within a single research project in order to understand a research problem more completely.[43] Quantitative research has been employed, which generates data that is either collected in numeric format or converted to it, while collated qualitative research tends to generate data that is largely retained in textual format; where non-textual data is collected, it is often expressed in textual format. With an explicit focus on the link between approaches, and the use of more than one theoretical position in relation to the data, the mixed-methods approach adopts a pragmatic position and brings together methods from the two major paradigmatic positions (positivist and constructive), conventionally regarded by some as incompatible.[44] This use of more than one approach is sometimes referred to as 'triangulation'.[45]

The research employed predominantly draws on the literature review and a location-based intercept survey designed to gather information on consumer behaviour, preferences, motivation or perceptions. In addition to these central methods, two key stakeholder interviews were conducted to provide an additional context. The triangulation of survey data, interviews and the literature review offered the prospect of enhanced confidence in the findings. For the purposes of primary research, an intercept survey was carried out in the town of Dachau in order to profile visitors. The survey provided a systematic method of collecting data from a sample of the population of interest. It utilised a semi-structured questionnaire which enabled a mix of quantitative and some qualitative information to be gathered. The design of the questionnaire included twenty-three questions concerning visitors' behaviour and motivation for visiting. The sample size consisted of 206 questionnaires and was aimed at visitors (day visitors and overnight) to Dachau. For the purpose of this study, a day visitor refers to a person who is not a resident in Dachau, and whose excursion to Dachau does not involve an overnight stay.

The survey was conducted throughout a number of locations in Dachau, including the Old Town, the Castle grounds, the Railway Station and the KZ entry point. The survey was conducted over a four-day period. A small team of volunteers were recruited and trained to conduct the survey in both English and German. When conducting the research a systematic sampling approach was adopted by which every third person was questioned. A card was handed out to participants containing information about the research and contact details in case participants wished to obtain findings of the study. The survey data was processed through the forms processing application TeleForm. Two key stakeholder interviews were also conducted in relation to the research project. While the interviews were not intended to contribute to the research as a method of generating new or representative knowledge, they were valuable in terms of obtaining additional data and commentary relating to camp visitation and Dachau's marketing, which would not have been hitherto available. The key stakeholders represented Culture, Tourism and Contemporary History, Stadt Dachau (Amt für Kultur, Tourismus und Zeitgeschichte) and the research services of Dachau KZ memorial site.

A mixed-methods approach offered improved accuracy in terms of the validation of findings and data was compared and analysed in terms of validity and accuracy. Traingulating in this way provided the research team with a wider perspective which served to enhance the completeness of the results and findings.[46] Thus, for example, the academic discourse on dark tourism to date has been largely based on qualitative approaches.[47] In order to produce a piece of research that will add to the literature, it is appropriate to adopt a quantitative approach to ascertaining visitor motivation in combination with other methods.

A limited timeframe for the project meant that the intercept survey was conducted on specific days within the year and at a specific time of year. All questionnaires were captured within a four-day period (Saturday to Tuesday), not including Wednesday to Friday or bank holidays. Furthermore, the survey was conducted in September, coinciding with the annual Oktoberfest in Munich. Consequently, this intercept survey can only claim to capture an impression of tourist visitation to Dachau. Furthermore, as this was a small-scale pilot study, the findings that can be extrapolated from the sample size are limited and can only suggest possible trends and indicators. In addition, the survey was limited by the language abilities of the research team, who surveyed in both English and German, but who were unable to reach the many non-English and non-German speaking visitors. The survey found that for the majority of visitors to Dachau, the main reason for visiting is the KZ memorial site. Some 68.9 per cent of visitors to Dachau visited the memorial site (whether it was the main reason for their visit or not), while 26.2 per cent visited the Old Town. Table 2.1 provides a breakdown of the attractions and sites within the Old Town and their visitor statistics. While the Old Town and the Castle seem to be the most popular town centre attractions, the KZ memorial site remains the most visited site in Dachau.

In terms of visitor numbers to individual Dachau sites over the course of a year, the KZ representative (2011) noted that the 2007 survey of the KZ 'estimated that there were about 700,000 to 800,000 visitors each year [. . .] I think that we have more visitors now than in 2007'. Therefore, a conservative annual estimate of visitor numbers to the KZ is circa 800,000 visitors per annum. In comparison, the most recent visitor data shows that visitor numbers to other attractions, such as the Painting Gallery (9,938 visitors), the District Museum (7,164) and the New Gallery (2,082), are significantly lower.[48] The survey undertaken showed that of the total number of visitors to the KZ, far fewer came from Munich or another German city than from outside Germany. It can be concluded that visitors from outside Germany

Table 2.1 Old Town and KZ tourist numbers

Site	Total no of visitors	Percentage
KZ Memorial Site	142	68.9
KZ only	111	53.88
KZ main reason for visiting	130	63.1
KZ + Old Town main reason for visiting	5	2.4
Old Town	54	26.2
Castle	47	22.8
Castle Gardens	27	13.1
St Jacob Church	17	8.3
City Hall	11	5.3
District Museum	8	3.9
Painting Gallery	6	2.9
Water Tower	3	1.5

(81 per cent) are significantly more likely to be visiting the KZ memorial site than visitors from Munich and other German cities (19 per cent).

German and non-German visitors

Table 2.2 provides a breakdown of German and non-German tourist visitors to the KZ memorial site and the Old Town. These figures show that when excluding German tourists from the visitor statistics for the KZ, the percentage of visitors to the memorial site is significantly higher. Some 92 per cent of non-German tourists visited the KZ memorial site, while only 33.3 per cent of German tourists visited the site. It can also be noted that 84.7 per cent of non-German tourists visited the KZ exclusively, while only 15.3 per cent of German visitors came for the memorial site only. The German tourists appear to have visited Dachau for a range of reasons, including visiting the Old Town (39.5 per cent) and related sites such as the Castle (37 per cent) and St Jacob's Church (14.8 per cent). German tourists also gave other reasons for coming to Dachau, such as visiting family and friends (7.8 per cent) and being in the area as part of an outdoor activity (8.3 per cent). Some 66.7 per cent of German visitors did not visit the KZ at all. It is notable that 84.9 per cent of German visitors did not put the KZ as their main reason for visiting Dachau, or did not visit the KZ at all.

The survey suggests that there is a trend among non-German tourists to specifically visit Dachau for the KZ memorial site, while German tourists come to Dachau predominantly for other reasons (although, as noted above, Germans are still the nationality most likely to visit the KZ).[49] It is also possible that the 15.3 per cent of Germans who came for the KZ exclusively predominantly comprised pupils and teachers who were visiting the KZ as part of an education programme. This is supported by data from the KZ research stakeholder interview which highlighted the pedagogical purpose of the site.

For 82 per cent of the total visitors to the KZ, it was their first visit to Dachau. The survey results suggested that most people only visit the KZ once. Of the total

Table 2.2 German and non-German tourist visitors

Statistically significant	Site	German tourists				Non-German tourists			
		No of visitors		Percentage		No of visitors		Percentage	
Yes	KZ Memorial Site	Yes	27	33.33		Yes	115	92	
		No	54	66.66		No	10	8	
Yes	KZ only	Yes	17	15.31		Yes	94	84.68	
Yes	KZ main reason for visiting	Yes	18	22.22		Yes	112	89.6	
No	KZ + Old Town main reason for visiting	Yes	1	1.23		Yes	4	3.2	
Yes	Old Town	Yes	32	39.50		Yes	22	17.6	
		No	49	60.49		No	103	82.4	
Yes	Castle	Yes	30	37.03		Yes	17	13.6	
		No	51	62.96		No	108	86.4	
No	Castle Gardens	Yes	18	22.22		Yes	9	7.2	
		No	63	77.77		No	116	92.8	
No	St Jacob Church	Yes	12	14.81		Yes	5	4	
		No	69	85.18		No	120	96	
No	City Hall	Yes	8	9.87		Yes	3	2.4	
		No	73	90.12		No	122	97.6	
No	District Museum	Yes	5	6.17		Yes	3	2.4	
		No	76	93.82		No	122	97.6	
No	Painting Gallery	Yes	4	4.93		Yes	2	1.6	
		No	77	95.06		No	123	98.4	
No	Water Tower	Yes	2	2.46		Yes	1	0.8	
		No	79	97.53		No	124	99.2	

first-time visitors to Dachau, 84 per cent visited the KZ and 79 per cent stated that this was their main reason for visiting. This indicates that first-time visitors usually visit Dachau for the KZ, which supports our first hypothesis. However, for residents of Munich the reverse was true: 17 per cent were first-time visitors, while 83 per cent were not.

The survey also found that while a significant majority of visitors to Dachau stayed overnight in Munich, only a minority stayed in Dachau (or elsewhere), and 27.2 per cent were day-trippers only (their entire trip was completed within one day and therefore did not include an overnight stay). Munich was the major location for overnight expenditure, which suggests Dachau was a day visitor expenditure

The statistics also suggest that Dachau benefits from its close proximity to Munich rather than being disadvantaged, in that tourists staying in Munich are likely to visit Dachau for the day, specifically to see the KZ memorial site. However, accommodation spend clearly gravitates to Munich. The number of visitors to

Dachau who were staying in Munich might have been highlighted in this survey because it took place during the Oktoberfest; indeed, a number of participants commented that they were visiting Dachau as a break from visiting the Oktoberfest, and it would appear that Dachau specifically attracts visitors who are normally resident in Munich, with their close proximity to the town meaning they are able to explore Dachau's cultural sites.

Marketing Dachau

It appears that most tourists do not use any of the information sources on offer (see Table 2.3), including Dachau's information brochures, which are available in the tourist information and memorial site information centres. However, 84.1 per cent of those who read the brochures found them helpful, suggesting that this particular current marketing channel is meeting the needs of the tourists who read them. Furthermore, as the tourist information centre is located in the Old Town, only the tourists who visited the Old Town would have had the opportunity to go to the tourist information centre.

At the time of the survey only 20.5 per cent of respondents had visited Dachau's official website, and when 'Dachau' was entered into a Google search, the official websites for the KZ and even the Youth Hostel were listed before the official town website. This indicates that the official town tourism website does not achieve high optimisation on the search engine and it is necessary to search for 'Dachau town' or 'Stadt Dachau', which reinforces the volume of associations on line between the name 'Dachau', the KZ and the events of World War II.

Some survey respondents' comments on the town merit consideration. For example, several German respondents expressed enthusiasm for the Old Town with its attractive architecture and ambience, highlighting its 'great way of life and culture'. Dachau's historic town was also described as 'an ideal destination for a day trip', and it is interesting to note that a very high proportion of tourists surveyed (86.7 per cent) were positive about their experience of Dachau. The proportion that gave the historical significance of the place as their reason for recommending Dachau (61.6 per cent) appears to correspond with the percentage of visitors whose main reason for visiting Dachau was the KZ (63.1 per cent). This would suggest that these visitors might have interpreted the question 'would you

Table 2.3 Dachau's marketing channels

	Have you visited Dachau's official website?	Have you visited Dachau's official YouTube channel?	Have you visited Dachau's tourist information centre?	Have you read any of Dachau's official information brochures?	If yes, did you find them helpful?
Yes	20.5%	1.5%	14.6%	19.9%	84.1%
No	79.5%	98.5%	85.4%	80.1%	15.9%

recommend Dachau' as 'would you recommend the KZ memorial site', and have highlighted its historical and educational value. Consequently, it is likely that when they return to their home countries they will refer to the KZ as 'Dachau', perpetuating this blurring of perception among non-German nationals.

As a tourist destination, Dachau's complex identity remains, and its international reputation as the place of the KZ makes it hard to successfully market an 'other Dachau'. Instead, Dachau lives in the shadow of its recent past. This is evidenced by the fact that visitor numbers to the KZ have consistently remained significantly higher than visitor numbers to the Old Town, despite its long history and previous identity as a centre for art and culture. It is clear that the majority of tourists visit Dachau primarily because of the KZ. However, it is interesting that the German tourists surveyed were in fact significantly more likely to be visiting the Old Town than the KZ. Furthermore, most of those who visited Dachau's other sites (non-KZ) were German tourists.

A study of this size and scope is invariably limited in its capacity to make any definitive recommendations, and further research is required to provide more comprehensive conclusions. However, there are some findings of interest. The impact of World War II's events on the external perception of Dachau as the site of a KZ remains deep-rooted more than sixty years after the end of World War II and the liberation of the KZ, a dark history which continues to cast a long shadow over the town. Of course, marketing of this destination's other features will always be challenged by its historically tragic reputation and shared location title, yet as an alternative to the tourist products of Munich, the KZ maintains significant appeal that shows no evidence of decline. The nature of the KZ tour's commodification as part of an urban break/event-related visit also merits highlighting and evidences the extent to which such sites are increasingly a core part of the visitor experience.

Notes

1 B. Distel et al., eds, *The Dachau Concentration Camp 1993–1945*, 4th ed. (Munich: Lipp Verlagsgesellschaft, 2005).
2 KZ Gedenkstätte Dachau, "Official Website," accessed January 7, 2012, www.kz-gedenkstaette-dachau.de/
3 KZ Gedenkstätte Dachau, *Studie* 2007; Sierp, 2011.
4 D. Riedel, Correspondence with the author, February 24, 2015.
5 P. Stone and R. Sharpley, eds, *The Darker Side of Travel: The Theory and Practice of Dark Tourism* (Bristol: Channel View, 2009).
6 J. J. Lennon and M. Foley, *Dark Tourism: The Attraction of Death and Disaster* (London: Continuum, 2000).
7 C. Rojek, *Ways of Escape* (Basingstoke: Macmillan, 1993).
8 Ibid., 136.
9 A. V. Seaton, "Guided by the Dark: From Thanatopsis to Thanatourism," *International Journal of Heritage Studies* 2 (1996): 234–44.
10 C. Wight and J. J. Lennon, "Towards an Understanding of Visitor Perceptions of Dark Attractions: The Case of the Imperial War Museum of the North, Manchester," *Journal of Hospitality and Tourism* 2, no. 2 (2004): 105–22.
11 P. Williams, "Witnessing Genocide: Vigilance and Remembrance at Tuol Sleng and Choeng Ek," *Holocaust and Genocide Studies* 18, no. 2 (2004): 234–54.

12 C. Rojek and J. Urry, *Touring Cultures: Transformations of Travel and Theory* (London: Routledge, 1997).
13 G. J. Ashworth and J. E. Tunbridge, "Old Cities, New Pasts: Heritage Planning in Selected Cities of Central Europe," *Geojournal* 49 (1999): 105–16.
14 G. Ashworth, "Holocaust Tourism: the Experience of Krakow," *Kazimierz. International Research in Geographical and Environmental Education* 11, no. 4 (2002): 363–7.
15 I. Wollaston, "Negotiating the Marketplace: The Role(s) of Holocaust Museums Today," *Journal of Modern Jewish Studies* 4, no. 1 (2005): 63–80.
16 J. Beech, "The Enigma of Holocaust Sites as Tourism Attractions: The Case of Buchenwald," *Managing Leisure* 5, no. 1 (2000): 29–41.
17 P. Stone, "A Dark Tourism Spectrum: Towards a Typology of Death and Macabre Related Tourist Sites, Attractions and Exhibitions," *Tourism: An Interdisciplinary International Journal* 52, no. 2 (2006): 145–60.
18 Lennon and Foley, *Dark Tourism*.
19 P. Stone, "Dark Tourism Spectrum."
20 H. G. Richardi, E. Philipp, and M. Lücking, *Dachau – A Guide to Its Contemporary History* (Dachau: Office of Cultural Affairs, Tourism and Contemporary History, 2001).
21 H. Marcuse, "Reshaping Dachau for Visitors: 1933–2000," in *Horror and Human Tragedy Revisited: The Management of Sites and Atrocities for Tourism*, ed. G. Ashworth and R. Hartmann (New York: Cognizant, 2005), 118–48.
22 Ibid., 119.
23 Ibid.
24 M. Fulbrook, *German National Identity after the Holocaust* (Cambridge: Polity Press, 1999).
25 H. Marcuse, *Legacies of Dachau: The Uses and Abuses of a Concentration Camp, 1933–2001* (Cambridge: Cambridge University Press, 2001).
26 Distel et al., *Dachau Concentration Camp*, 26–7.
27 Ibid., 27.
28 Ibid.
29 Fulbrook, *German National Identity*.
30 Marcuse, *Legacies of Dachau*, 139.
31 *KZ Gedenkstätte Dachau*.
32 *Empirische Analyse der Besucher der KZ Gedenkstätte* (Dachau: Sierp, 2011).
33 Distel et al., *Dachau Concentration Camp*.
34 *Empirische Analyse*.
35 Distel et al., *Dachau Concentration Camp*.
36 Richardi et al., *Dachau*.
37 Stadt Dachau, "Imagefilm," accessed September 17, 2011, www.youtube.com/user/StadtDachau.
38 *Empirische Analyse*.
39 K. Dinnie, *City Branding: Theory and Cases* (London: Palgrave, 2011); J. Heeley, *Inside City Tourism: A European Perspective* (Bristol: Channel View, 2011); J. A. Mazanec, ed., *International City Tourism* (London: Pinter, 1997).
40 B. Baker, *Destination Branding for Small Cities: The Essentials for Successful Place Branding* (Oregon: Creative Leap, 2007).
41 B. M. Kolb, *Tourism Marketing for Cities and Towns: Using Branding and Events to Attract Tourism* (London: Elsevier/Butterworth-Heinemann, 2006).
42 R. Bennett and S. Savani, "The Rebranding of City Places: An International Comparative Investigation," *International Public Management Review* 4, no. 2 (2003): 70–87.
43 M. Denscombe, *The Good Research Guide: For Small-scale Social Research Projects* (Milton Keynes: Open University Press, 2010); J. W. Creswell, *Educational Research: Planning, Conducting, and Evaluating Quantitative and Qualitative Approaches to Research* (Upper Saddle River, NJ: Merrill/Pearson Education, 2002).

44 A. Armitage, "Mutual Research Designs: Redefining Mixed Methods Research Design" (paper presented at the British Educational Research Association Annual Conference, Institute of Education, University of London, September 5–8, 2007).
45 A. Bryman, *Social Research Methods* (Oxford: Oxford University Press, 2008).
46 Denscombe, *Good Research Guide*.
47 P. Stone and R. Sharpley, "Consuming Dark Tourism: A Thanatological Perspective," *Annals of Tourism Research* 35 (2008): 574–95.
48 J. Mannes, Visitor numbers of Dachau Painting Gallery, District Museum and New Gallery 2010 (personal email communication, October 11, 2011).
49 A. Sierp, "Remembering to Forget? Memory and Democracy in Italy and Germany" (paper prepared for the XXIII Convegno SISP, Facoltà di Scienze Politische LUISS Guido Carli, Roma, September 17–19, 2009).

3 Prison tourism
Exploring the spectacle of punishment in the UK

Sarah Hodgkinson and Diane Urquhart

Introduction

People have always had a somewhat morbid and voyeuristic fascination with the macabre: a desire to witness and vicariously experience violence, suffering and death. For most of us, with no first-hand experience of criminality and punishment, prison tourism offers us the opportunity to see inside these 'closed' institutions, to experience a range of motivations and become part of a growing interest in the 'spectacle of punishment'.[1] In this way, visits to institutions of punishment become the ultimate voyeuristic criminological experience. Tourism to such sites, in the form of decommissioned prisons, prison museums and carceral tours, is, however, hugely controversial. In this chapter we will briefly review the growing research in this field and discuss the UK context, where penal tourism is less developed and the sites relatively unexplored. In our discussion of two historical sites – the Galleries of Justice in Nottingham and Derby Gaol – we will consider the commodification of prisons for public consumption, and ask whether such institutions present a sanitised, sensationalised or distorted view of offenders and their punishment.

Setting the scene

Throughout history the human desire to witness the suffering, and even death, of others is well documented. While the allure is arguably as old as history itself, academic interest has grown over recent years with the emergence of research into 'dark tourism', defined by Richard Sharpley as 'travel to places associated with death, disaster and destruction'.[2] While this term was derived from tourism studies, its significance has been recognised within various other disciplines (e.g. history, museum studies, geography, criminology and psychology). More importantly, the concept has been applied to a multitude of potentially 'dark' sites and become a term that has filtered down into public consciousness, with numerous websites and blogs devoted to the subject. As well as the growing scholarly and media attention dedicated to dark tourism, literature suggests both heightened demand from the public[3] and increasing availability of 'dark' sites.[4]

The crossover between 'dark tourism' and sites of criminological significance has been recently acknowledged.[5] As most of us will have no first-hand experience of serious crime, we rely upon public sources of information for our knowledge, from what we see or read about in the media to the places we visit during our recreational time.[6] In this way, we satisfy our voyeuristic desires to seek out more 'authentic' experiences such as those offered by tourism.[7] John Lennon notes the growing demand and supply of crime-related tourism in modern popular culture, making this once quite 'niche' experience now a mainstream way of 'experiencing' crime.[8] Undoubtedly the term 'dark tourism' has negative connotations, implying that our interest is born out of a desire to be entertained by the suffering of others. In addition to satisfying our curiosity about aspects of life beyond our own experiences, crime-related dark tourism can potentially provide us with the reflective space to learn about, witness and commemorate the death, suffering and punishment of others.[9] Prison tourism has begun to be explored in terms of the (mis)use of narratives of the carceral past to educate through the use of prison museums and carceral tours.[10] Perhaps it is a feature of modernity to look to these public retellings of penal histories as contemporary 'outsiders'; seeking to explore what remains firmly outside our own experience, as the sanitisation of punishment provides limited opportunity to witness it directly.

There is some credence to the idea that the invisibility of punishment in society today fuels the public's fascination with it.[11] To the vast majority the prison is closed and distant, despite its often close proximity to the local community. Paradoxically, the prison is a visible (if not entirely credible) deterrent. It is a public institution, and yet punishment itself remains a very private practice. This sits in stark contrast to the spectacle of punishment of former days, where Michel Foucault made explicit the symbolic and ritualistic importance of the body as a site of punishment and torture.[12] His well-known narrative charts the supposed decline of 'Punishment-as-Spectacle', as the public execution retreated behind the secret walls of the prison.[13] Although John Pratt related this development to a civilising process that evolved from Enlightenment thinking, this is a potentially idealist interpretation.[14] While physically out of sight, execution remained very much in people's psyche, and the crowds would still flock to witness a noteworthy execution.[15] The privatisation of the execution did not quell the desire of a curious public to be involved with the proceedings; their interest remained – and remains still.

This ongoing quest to witness punishment is today reflected in intense media coverage, with an emphasis on the extremes of crime such as violence and homicide, as well as punitive responses to those who are seen to transgress our moral and legal boundaries.[16] While the media raises our awareness of crime and punishment, it leaves us de-sensitised to it all the same.[17] It could therefore be argued that in this demand for crime-related entertainment the media spectacle is no longer sufficient, especially when travel is becoming increasingly accessible and offers the opportunity to visit the scenes of crime and punishment ourselves. Although research evidences the increased supply of such dark tourist sites, much less is known about the demand for these sites in terms of the motivations of 'dark tourists'.[18] When considering the broad variety of penal sites, visitor motivations may well depend upon the type of experience on offer. Prison tourism sites span the whole spectrum of dark tourism, from those that are very much 'out of authentic space' to those sites with a rich and long penal history. Some are much 'lighter' than others, with the emphasis on easy shocks and vicarious thrills, such as Stone's titillating 'fun factories' in the London, York and Edinburgh dungeons.[19] However, some are 'darker' sites of mass incarceration, suffering and execution, such as Tuol Sleng, the former detention centre of the Khmer Rouge in Cambodia.[20]

Considering two English sites, we discuss how carceral histories are sometimes recreated and re-imagined, as well as how they manage to combine historical and educative authenticity with an entertaining tourist experience. The presentation and reconstruction of the penal past for public consumption is often controversial, as sites try to cater to often macabre public demand while trying to preserve historical buildings. Consequently, exhibits and educational tours are more often centred on gory tales or gruelling punishments in order to generate appeal with a focus on death, to capitalise on the potentially voyeuristic nature of the visitor. Vic Gatrell considers the historic practice of bearing witness to a public execution and suggests that the motivation to view the deed was primal, and satisfied some basic hidden curiosity.[21] Kevin Walby and Justin Piché note how artefacts of punishment are exhibited in such a way that is reflective of practices rooted firmly in a primitive history.[22] As such, it could be suggested that these sites can satisfy some hidden basic instinct, while the visitor is also reminded that punishment today has much improved. In this sense such heritage sites can be seen as 'conduits between the past and the present', combining 'lived experience with myth'.[23]

The penal tourist experience can thus be seen as an important means of shaping our attitudes and understandings of prisons and prisoners today. John Lennon proposes that current-day justifications for visiting 'dark sites' are about gaining knowledge.[24] Furthermore, the interpretations and understandings gleaned from attending penal sites can induce reflection upon both historical and contemporary approaches to punishment.[25] Michelle Brown makes an important observation in this respect when she suggests that the practice of observing punishment can mean that it becomes quite customary: 'In its most dangerous manifestations, spectatorship normalises what should be abhorred.'[26] There is potential, therefore, for visitors to become desensitised to representations of punishment both within and outside the penal tourist site, somewhat reminiscent of a former (and less civilised) society. This can serve to invoke a more defined social distance from those who face punishment today, re-emphasising their status as deviant 'other'.[27]

Opportunities to see inside functioning prisons are rare, and are usually restricted to carceral tours for educative means. As their purpose is academic rather than voyeuristic, there is debate about whether this should be discussed within the context of 'dark tourism'.[28] In practice, however, although the aim may be to educate about life inside our prisons, these tours are inevitably short and highly controlled, and visitors may well still approach them with voyeuristic tendencies. The educative utility and morality of carceral tours is therefore heavily contested. These staff-led tours tend to over-simplify and sanitise the realities of punishment, and reveal little about prisoners themselves, who are typically excluded or marginalised within these experiences.[29] Indeed, prisoners report feeling like a 'specimen on display',[30] exploited for public consumption and judgement. Such visits can therefore serve to further reinforce negative stereotypes about 'dangerous offenders' and legitimise the need for incarceration.[31] However, others argue that, if done ethically and reflectively, they also have the ability to challenge such stereotypes and question modern punishment practices.[32]

Prison tourism in the UK

Although there is growing international literature on prison tourism, there is very little research to date exploring this in the UK. This is despite several very interesting sites that span Philip Stone's spectrum of dark tourism, from those at the lighter entertainment-focused end of the spectrum to more authentic and historically factual experiences.[33] There is also considerable crossover with 'ghost tourism', as more prison tourist sites choose to market themselves in this way to remain sustainable and attract new audiences.[34]

For this research we visited a number of prominent sites within England. At these sites we spoke to curators and site managers and took part in organised tours, as well as informal tours 'behind the scenes'. We were interested in finding out about the development of these sites, the demand for these experiences and witnessing the tourist experience first-hand. For the purposes of our research we focused on two sites: the Galleries of Justice in Nottingham and Derby Gaol. We felt that these sites offered something quite unique in the history of imprisonment: a period that represents one of the 'darkest' times, charting the stories of Georgian and Victorian systems of incarceration and capital punishment. Additionally, there is only limited historical information available within the public domain on these sites, and they have received relatively little interest.

Prison tourism is somewhat undeveloped in the UK when compared to the USA, Canada and Australia, both in terms of tourist opportunities and academic interest. For the general public there are limited opportunities to experience our carceral history as a tourist; this is especially so for those interested in our more recent penal past, although carceral tours are available in the UK to certain groups. For example, the more 'experimental' prisons, such as HMP Grendon Underwood, have regular 'social afternoons' where members of the public can apply for a chance to mingle with the inmates.[35] As discussed above, such events and carceral tours are, in themselves, controversial, as they can be said to patronise the inmates and feed our desires as 'outsiders' viewing 'insiders',[36] although they can also potentially challenge prevailing stereotypes.[37] Alana Barton and Alyson Brown discuss the persistent tourist appeal of HMP Dartmoor, one of England's oldest prisons still in use today.[38] Whereas in the past visitors clamoured to scale the walls or crowd around the entrances, there is now a prison museum on site to welcome tourists and sell souvenirs. Glimpses of actual prison life are, however, not sanctioned, and opportunities to see inside functioning prisons are rare within the UK.

Decommissioned prisons are therefore often the primary sites of prison tourism, and these are documented well in the international literature.[39] These large, iconic nineteenth-century penitentiaries do not, however, exist in the UK, where some buildings tend to have been demolished and the sites reused, while many other Victorian prisons remain in use. Other decommissioned prisons have been redeveloped for different functions while still capitalising on their penal history to provide an added attraction or marketing slant, a good example of which is Malmaison Oxford Hotel, based in the former HMP Oxford. Such sites thus reconstruct and re-imagine the carceral space for different purposes, while exploiting its history for profitable gain. Reuse of such sites and their public popularity reflects our fascination with the architecture of prisons; as Jacqueline Wilson notes, we are often drawn to the prison building rather than the lives of former inhabitants.[40] This raises ethical dilemmas about how we commodify the history of our prisons and neglect the narratives of those that suffered within the prison itself.

Closed prisons as tourist sites have therefore been dominated by prison museums located in prisons from our more distant history, typically the Georgian or Victorian local gaol (e.g. Ripon Prison Museum in Yorkshire; Ruthin and Beaumaris gaols in Wales). Again, very little has been written about prison museums in the UK. Michael Welch[41] compared the Clink Museum in London to prison museums in Brazil and Australia, and describes the conflict of portraying the prison as 'a morbid theme park' while also providing some valuable historical information to visitors.[42] He notes how at times this balance is well maintained, allowing visitors to reflect on issues about poverty and incarceration, our notions of justice and tolerance and the governance of the criminal population. He also recognises the tendency to sensationalise and focus on the most gruesome aspects of torture, suffering and pain. This is particularly problematic as the visitor is positioned as a voyeur, looking in at the lives of others from a safe distance.[43] Jennifer Turner and Kimberley Peters have also recently written about the Galleries of Justice in Nottingham, and similarly recognise the potential to create sensationalist representations and stereotyped characters to entertain.[44] As they conclude, 'As such, horror for entertainment prevails, and "true" horrors are sanitised'.[45]

Additionally, more entertainment-orientated or 'lighter' dark tourist experiences exist, sometimes on the site of an authentic prison, sometimes in a site quite unconnected with prisons. A prime example of these 'dark fun factories', which cater to a desire to be scared and entertained, are the Dungeons Attractions that operate in London, York and Edinburgh in the UK. As Stone suggests, these 'create a *safe* congregant space where *unsafe* ideas of the taboo may be inspected close up through a morbid gaze'.[46] These attractions reside firmly within mainstream commercial tourism and involve a focus on the darker tales and myths of local history, using the 'performative experience' of costumed actors, often with little historical accuracy or authenticity. They are an acknowledged commodification of the past.[47]

Prison tourist sites present the opportunity to move beyond our mediated constructions of crime and punishment; they facilitate the potential to witness and participate in a reconstruction of the past. Some of these experiences are very open in terms of their desire to thrill, shock and entertain; a spectacle of punishment fit for modernity.[48] Others strive to maintain an ethical and historical stance, without recognising the potential problems of commodification and sanitisation of the darkest periods of our historical past. In this sense we view the prisoner and their suffering at a comfortable distance,[49] and they remain safely beyond our moral universe.[50] Visitor expectations and experiences will fluctuate, however, depending on the nature and presentation of the site.

We explore two sites, both of which are located within former Georgian–Victorian local gaols. They fall within the darker end of Philip Stone's spectrum of dark tourism, in that they have clear educational purposes and reflect the brutal conditions and harsh regimes of the past (and were therefore places of considerable torment and suffering)[51] – although, as we shall see in the case studies presented below, they each also rely on sensationalism and the popular allure of the supernatural. Their tours and exhibits inevitably represent a commodified 'performance' of specific narratives of history. Both also strive to offer diverse interpretations of the site to attract differently motivated audiences.

Galleries of Justice

The Galleries of Justice in Nottingham incorporates the former Shire Hall, Nottinghamshire County Gaol and the County Police headquarters. As a site of penal heritage, it has a long and varied history. Courtrooms were in use here from 1375, and the medieval prison cells and oubliette from 1449 remain. The building itself dates to the 1600s, although it was expanded in the 1830s.[52] The prison continued to be used into the Victorian era, slowly implementing some of the reforms that were taking place nationally, but was eventually closed in 1878 after the Prison Act 1877 nationalised the prison service. Some parts of the site (the courtrooms and police station), however, remained in use until 1986. Derelict for almost ten years, the site was purchased and opened as the Galleries of Justice in 1995 (see Figure 3.1). Current estimates suggest that there are approximately 40,000 visitors per year, and numbers are increasing.

We visited the Galleries of Justice several times from January to June 2015. During this time we took part in the standard tours, as well as attending two of their late-night paranormal events. We also conducted interviews with their Senior Curator/Archivist and resident spiritualist medium.[53] In addition, we were fortunate to be allowed access to parts of the site not generally open to visitors, such as the medieval cave cells and tunnels underneath the prison, and the adjoining disused police station.

The Galleries of Justice has considerable historical significance, charting the history of punishment from medieval times, to the tumultuous reforms of the 1780s, through to the prison's closure in 1878. The museum allows visitors to reflect upon the modern development of prisons and our attitudes to incarceration. For example, since 2005 it has housed the National Prison Service Collection, and parts of the site remained in use until the 1980s. It can therefore tell the history of prisons from local gaols through to our current prison system. Prior to the nineteenth century, places of incarceration were primarily about confinement until trial, execution or the resolution of debts, and not about prison as punishment *per se*.[54] The site therefore documents the decline of the local gaol and the 'birth' of the 'modern prison'. It has an important place in the history of prison reform, featuring

Figure 3.1 The Galleries of Justice, Nottingham, incorporating Nottinghamshire County Jail.

Photo copyright S. Hodgkinson.

in John Howard's (1777) *The state of the prisons in England and Wales* as one of the worst prisons in England, second only to Newgate Gaol in London.[55] It also illustrates how subsequent reforms were often never actually realised within these local prisons.[56] Subsequently, it was one of the 122 prisons closed after the nationalisation of the prison service in 1877. Furthermore, with 2,000 prisoners passing through the prison bound for Australia, it additionally documents the history of transportation, and since it was also instrumental in the abolition of public execution in 1868,[57] it focuses on the barbarous conditions of Georgian and Victorian gaols and the 'growing distaste for the carnival of public execution'.[58] It presents rich narratives about this dark period in England's prisons as well as penal reform during this pivotal point in history.[59] Encompassing a strong sense of location authenticity, the site today remains relatively unchanged since its Georgian and Victorian period of use.[60]

The Galleries of Justice offer a variety of ways in which the visitor can experience the site. The standard daytime Crime and Punishment tour involves visitors being met at three different points by costumed interpreters who perform the roles of key courtroom and prison staff. The aim is to impart a sense of what life was like in the prison and the broader historical changes that were taking place, but in an entertaining way. As the Senior Curator told us:

> We use costumed interpreters to help navigate the visitors around the site . . . It's informative, it's entertaining, it's educational. The thing being that the

general visitors don't realise they're being educated from that because it's done in such an engaging and entertaining way.

Parts of the site are also left for the visitor to explore on their own, and there are various exhibitions and prisoner biographies throughout the site, as well as props to help them visualise what life was like as a prisoner.[61] The use of costumed interpreters and role-play (on the part of both the tour guide and the visitors) is the subject of some debate. As Michael Welch found in his visit to The Clink in London, the entertainment and 'fun' can be balanced with accurate historical content, facilitating the visitor to learn and reflect upon these 'darker' aspects of punishment within their historical context.[62] It also tries to instil some reflection upon the people who resided within the walls of the prison (both prisoners and staff), rather than just the history and architecture of the prison, which can be portrayed as almost devoid of human context.[63] But it is also reminiscent of the lighter 'fun factory' experience, where much is made of the brutal conditions, torture and executions, to entertain visitors and capture their morbid imaginations.[64] In this way the costumed tour is very much 'theatre'; it is a performance for public consumption of the more 'exciting' elements of Georgian and Victorian penality. As a 'penal tourist' we therefore remain at a distinct distance from these historical events and characters, which can serve to provide us with a sense of relief,[65] preventing us from having to face the true extent of the inhumane treatment of the prisoners.[66] As Alana Barton and Alyson Brown suggest:

> It provides a dramatic space whereby one group of people (the 'law-abiding audience'/'us') can experience the world of the 'other' (the 'criminal actors'/'them') whilst, at the same time, remaining untainted by the ignominy and denunciation that normally defines both the domain and its denizens.[67]

In an attempt to diversify and appeal to different audiences, the Galleries of Justice has regularly opened on evenings since 2008. Every Friday evening Ghost Tours are held and visitors are guided by a costumed interpreter, with a focus on the 'haunted' landmarks. On Saturday evenings the 'Terror Tours' take place, which have an even greater entertainment focus, featuring actors in costumes jumping out at visitors during the tour. In recent years the Galleries of Justice has seen a considerable expansion in terms of what it offers to sustain and expand its profitability. For example, it increasingly hires out the site as a venue to a range of external companies for paranormal investigations. These representations of the site in particular re-imagine the site and its history with little concern for historical context, and are seen as a chance to make pure profit on the part of the museum.[68] As Stone suggests, such 'narratives of fear and the taboo are extracted and packaged up as fun, amusement and entertainment and, ultimately, exploited for mercantile advantage ... an increasingly integral component of (dark) tourism marketing'.[69] This tendency to market the site to appeal to different audiences is also apparent in our second case study, Derby Gaol, where there is arguably an even more pronounced drive towards satisfying visitors' more macabre motivations.

Derby Gaol

Situated in Friar Gate, Derby Gaol is a much smaller and more intimate venue than the Galleries of Justice and lacks the more business-centred approach favoured by the latter. This is potentially due to the gaol being a private concern; it is currently owned by celebrity TV historian and 'ghost-hunter' Richard Felix, who manages the site with the help of a small staff. During our visits to the gaol, we were fortunate to have the current curator spend some time with us personally, and to speak with Richard Felix.

Felix purchased the site in 1997 and it was featured in the popular TV show *Most Haunted*, in which he was cast as the resident historian. Parts of the gaol have since been restored by the owner to what is thought to be its original state.[70] Thus, and in contrast to the Galleries of Justice, the way in which Derby Gaol is presented is reflective of the interests, beliefs and passions of its owner, and not merely as a commercial enterprise. Visitors to the site are therefore primarily those who share these supernatural interests, with a particular focus on the gaol's purported reputation of being haunted. Although the site is a smaller and much less publicised venue than the Galleries of Justice, it is estimated that between four and five thousand visitors experience Derby Gaol each year, with the most popular events by far being those which capitalise on the site's haunted heritage.

Built in 1756, Derby Gaol was used to house all manner of criminals awaiting trial or execution, as well as local debtors who, if fortunate, were held separately from offenders in an open common room. Originally designed to hold twenty-one prisoners and twenty-six debtors, the gaol frequently held more, and in 1819 it reportedly held sixty-nine prisoners in extremely squalid conditions. The seven small cells were reserved for the very worst offenders, although those with the necessary means could buy themselves a little comfort at the profit of the gaol-keeper.[71] Today two cells remain, one of which was assigned to the condemned, and each have their original doors intact and inscribed with the graffiti of former captives. What remains of the site is small, but its atmosphere and rustic authenticity more than compensate for its lack in size. While officially a museum, the gaol is only open to the public on Saturdays, but offers a variety of 'attractions' to entice the visitor. In addition to providing the visitor with the opportunity to inspect the remaining cells, the site is home to the Derby Police Museum, which opened in July 2014; the wide range of exhibits includes uniforms, truncheons and nineteenth-century handcuffs.

As has been discussed in the case of the Galleries of Justice, there is the need for sites to diversify and adapt to public demand and interest. Currently, Derby Gaol provides educational tours to schoolchildren on the theme of crime and punishment, but it also makes the best use of its status as a site of punishment, confinement and death. Furthermore, following the site's appearance on UK television, there is understandably much reference made to its reputation as being haunted. Two 'ghost walks' are now offered by Derby Gaol: one which tours the reputedly haunted venues within Derby itself and another which includes the gaol and surrounding area of Friar Gate, 'The Hangman's Walk' (those on the latter are invited back to the gaol for a 'condemned man's supper'). The gaol also hosts overnight vigils and is available for private hire to other 'ghost-hunting' groups, or to celebrate a special occasion in a 'spooky' way. It is also the current regular meeting place for the Derby Heritage Traditional Music Club.

The fate of those who passed through the gaol's doors was decided at St Mary's Gate courts, and here they stayed before being moved elsewhere, or more often (as befitted the Bloody Code) to the gallows. It is reported that at least fifty-eight people were hanged at Derby Gaol between 1756 and 1825, including two women.[72] Those who committed murder were also further punished after death; once hanged their bodies were taken to St Mary's Gate, where they would be dissected by surgeons in full view of the crowd.[73] William Hogarth's *The Reward of Cruelty* presents a graphic illustration of such a scene, where the body of highwayman Tom Nero meets a grisly end following execution at Tyburn.[74] Thus, the body of the condemned was used to further humiliate the perpetrator, with the spectacle presenting a form of deterrent.[75] Furthermore, the physical dissection of the body instilled a certain fear within the perpetrator that extended beyond death itself,[76] as Richard Felix explains:

> The body wasn't whole and that meant that on the day of judgement, no physical resurrection for you. And where are you going to go? Straight to Hell. Death wasn't the problem, death was commonplace ... it was the afterlife that had such an important role in everyday life.

A well-documented, famous and particularly gruesome execution that took place at the gaol was the hanging in 1817 of Brandreth, Turner and Ludlam, who led the Pentrich Rebellion. Vic Gatrell[77] reports that a crowd of six thousand flocked to Derby to witness the scene, and an article from *The Times* told how the scaffold was furnished with a block of wood, sacks of sawdust to catch the blood, two axes, two sharp knives and a basket to hold the heads.[78] At the sight of Brandreth's decapitation it was reported that 'There was a burst of horror from the crowd ... the instant the head was exhibited, there was a tremendous shriek set up, and they ran violently in all directions, as if under the impulse of sudden frenzy'.[79]

It is also noted, however, that some onlookers remained at the scene to quietly watch on as the heads of Brandreth's accomplices met the same fate. Replicas of the death masks taken from the heads of the men are displayed at Derby Gaol and present a particularly haunting sight (see Figure 3.2).

Notwithstanding its gory stake in English penal history, very little has been written about Derby Gaol as a site of punishment. Moreover, any focus on the gaol in the local or national media has viewed the history and significance of the site primarily from its reputation as a haunted location. This is despite its historical significance and the rarity of public access to prisons from this period of history. Reality is inextricably enmeshed with myth and a reliance on the ghostly retelling

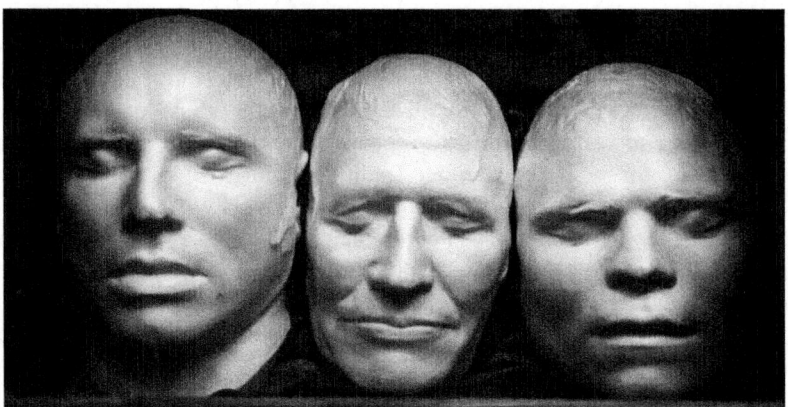

Figure 3.2 Death mask replicas of Brandreth and accomplices, Derby Gaol.
Photo copyright S. Hodgkinson.

of the site, and yet it is undoubtedly the continuation of such myth that sustains its existence. As a former site of confinement, death and suffering, Derby Gaol appears at the very darkest end of Phillip Stone's 'dark tourism spectrum', but also embraces the 'light' and 'lightest' aspects through the provision of ghost walks and as host to children's tours and a local music club.[80] What is of note, however, is that Derby Gaol retains a sharp focus on the historical aspects of the site, and the experiences on offer are marketed with due consideration of this; historical authenticity is firmly at its core.

Conclusion

Prison tourist sites are diverse, and range from those offering a somewhat superficial and sensationalist reconstruction of historical punishment to those highly sensitive to history, and offering more pluralistic interpretations. The increased supply and demand of such sites for the interested 'dark tourist' can be seen to reflect an ongoing interest in 'witnessing' crime and punishment for ourselves, beyond the mediated prison film.[81] In the search for an 'authentic' experience we visit sites of current or former prisons, and the artefacts, sites and lives of past prisoners become a spectacle of punishment commodified for public consumption. The penal gaze, rendered accessible through modern tourism practices,[82] further serves to accentuate the social distance between 'us' (i.e. the interested tourist) and 'them' (the prisoner, past or present).[83] Visits to functioning or decommissioned prisons, or to entertainment-focused attractions, can be regarded as the ultimate criminological voyeuristic experience. How we navigate and consume such sites, and the tours, exhibitions and artefacts within them, is of immense value to criminology. They allow us the reflective space to critically rethink notions of crime and punishment, and are open to a polysemy of constructions and interpretations.[84]

The use of such sites to educate and commemorate has been widely debated. While they are clearly more than sites of 'dark tourism' for the casual yet morbid voyeur, and cater to a range of visitor motivations, they must remain attractive to a variety of audiences in order to remain sustainable and allow site preservation. As we found in our own research, the majority of our decommissioned prisons remain in a derelict condition, have been demolished completely or have been redeveloped into recreational spaces quite disconnected from their historical use as a site of imprisonment and suffering. There are a host of ethical dilemmas concerning what can and should be done with these sites. For some sites this can be both politically and morally contentious in terms of what narratives of history they choose to represent, whose stories they tell or whether they are regenerated as sites of recreation with little regard for what the site once was. One would think these dilemmas diminish with sites located within our historical past, but dangers remain in terms of glorifying the pain and suffering of prisoners and implicitly accepting the practice of incarceration as a valid form of punishment.

In this chapter we discussed two sites in England: the Galleries of Justice in Nottingham and Derby Gaol. Both represent the decline of the local gaol, and were closed in the move towards more large-scale imprisonment with the nationalisation of the prison service from the 1870s. They focus on the brutal and inhumane conditions within Georgian and Victorian gaols, and the growth of the criminal population during these times. They also chart the process of reform, and provide a comparison in this sense between our systems and practices of imprisonment today, compared to our more 'uncivilised' past. In this sense, while giving visitors an entertaining but informative view of life inside these bygone prisons, they focus on how historically prisons were inhumane, ineffective and degrading while obscuring a critical reflection on how they may remain all of these things today. The Galleries of Justice makes great use of costumed interpreters to navigate visitors around the site and encourage role-play (the visitor as the prisoner and the tour guide as a figure of authority such as a warden or courtroom official). While on the one hand this facilitates a more active engagement with the site and its history, and focuses the visitor on those who were imprisoned here, this also risks creating stereotyped caricatures, whose crimes and punishments become a source of mirth and entertainment. It further serves to accentuate the power differential between the prisoner, whose suffering is perhaps trivialised, and the spectator.

Both sites play upon our morbid fascination with pain, torture and execution, and display 'props' and artefacts to capture the imagination of the tourist. They also both offer 'ghost tourism' to increase revenue, risking some exploitation of their history and a reconstructed and distorted presentation of the site. Some compromise between sustainability and historical authenticity is, however, probably inevitable at former penal sites such as these.[85] The rise of interest in the paranormal seems to have opened up new opportunities and audiences, which may help fund the preservation of sites for future generations to experience and explore.

More research is needed to further explore what meanings we can impart from often sanitised and dehumanised presentations of prisons, and also how more complex, contested and nuanced presentations of the sites can be encouraged.

While recognition is made that more plural constructions of punishment are possible,[86] the 'tourist gaze' all too often reifies difference and distance between the 'viewer' and the 'viewed'.[87] The constructed narratives about punishment can also facilitate popular punitivism, and an assumption that incarceration is both necessary and uncontested today. Too often the suffering of our past prisoners becomes a source of shallow entertainment, or imparts the idea that prisons, although once bad, are now somehow unproblematic.

Notes

1 M. Foucault, *Discipline and Punish: The Birth of the Prison* (London: Penguin, 1991).
2 R. Sharpley, "Shedding Light on Dark Tourism: An Introduction," in *The Darker Side of Travel: The Theory and Practice of Dark Tourism*, ed. R. Sharpley and P. R. Stone (Bristol: Channel View Publications, 2009), 9.
3 J. Lennon and M. Foley, *Dark Tourism* (Andover: Cengage Learning EMEA, 2010).
4 Sharpley, "Shedding Light."
5 J. Z. Wilson, *Prison: Cultural Memory and Dark Tourism* (Oxford: Peter Lang, 2008).
6 N. Rafter, *Shots in the Mirror: Crime Films and Society*, 2nd ed. (Oxford: Oxford University Press, 2006).
7 S. Hodgkinson, "The Concentration Camp as a Site of Dark Tourism," in *Témoigner: LEntre Histoire et Mémoire* 116 (2013): 22–33; D. Dalton, *Dark Tourism and Crime* (London: Routledge, 2015).
8 J. Lennon, "Dark Tourism and Sites of Crime," in *Tourism and Crime: Key Themes*, ed. D. Botterill and T. Jones (Oxford: Goodfellow, 2010), 215–29.
9 Dalton, *Dark Tourism and Crime*; B. T. Knudsen, "Thanatourism: Witnessing Difficult Pasts," *Tourist Studies* 11, no. 1 (2011): 55–72.
10 Wilson, *Prison*; M. Brown, *The Culture of Punishment: Prison, Society and Spectacle* (New York: New York University Press, 2009); Justin Piché and Kevin Walby, "Problematizing Carceral Tours," *British Journal of Criminology* 50 (2010): 570–81.
11 Brown, *The Culture of Punishment*.
12 Foucault, *Discipline and Punish*.
13 Ibid., 9.
14 J. Pratt, *Punishment and Civilization* (London: Sage Publications Ltd, 2005).
15 V. Gatrell, *The Hanging Tree: Execution and the English People 1770–1868* (Oxford: Oxford University Press, 1994).
16 M. O'Neill and L. Seal, *Transgressive Imaginations: Crime, Deviance and Culture* (Basingstoke: Palgrave, 2012).
17 Stanley Cohen, *States of Denial: Knowing about Atrocities and Suffering* (Cambridge: Polity Press, 2001).
18 Hodgkinson, "The Concentration Camp."
19 P. R. Stone, "A Dark Tourism Spectrum: Towards a Typology of Death and Macabre Related Tourist Sites, Attractions and Exhibitions," *Tourism* 54, no. 2 (2006): 145–60.
20 Dalton, *Dark Tourism and Crime*.
21 Gatrell, *The Hanging Tree*.
22 K. Walby and J. Piché, "The Polysemy of Punishment Memorialization: Dark Tourism and Ontario's Penal History Museums," *Punishment and Society* 13, no. 4 (2011): 451–72.
23 J. M. Rickly-Boyd, "The Tourist Narrative," *Tourist Studies* 9, no. 3 (2010): 259–80.
24 J. Lennon, "Dark Tourism and Sites of Crime."
25 C. Strange and M. Kempa, "Shades of Dark Tourism: Alcatraz and Robben Island," *Annals of Tourism Research* 30, no. 2 (2003): 386–405.
26 Brown, *The Culture of Punishment*, 197.

27 C. Greer and Y. Jewkes, "Extremes of Otherness: Media Images of Social Exclusion," *Social Justice* 32, no. 1 (2005): 20–31.
28 F. Pakes, "Howard, Pratt and Beyond: Assessing the Value of Carceral Tours as a Comparative Method," *The Howard Journal of Criminal Justice* 54, no. 3 (2015): 265–76.
29 Piché and Walby, "Problematizing Carceral Tours."
30 C. Minogue, "Human Rights and Life as an Attraction in a Correctional Theme Park," *Journal of Prisoners on Prisons* 12 (2003): 44–57.
31 Piché and Walby, "Problematizing Carceral Tours."
32 D. Wilson, R. Spina and J. E. Canaan, "In Praise of the Carceral Tour: Learning from the Grendon Experience," *The Howard Journal of Criminal Justice* 50, no. 4 (2011): 343–55.
33 Stone, "A Dark Tourism Spectrum."
34 S. Hodgkinson and D. Urquhart, "Ghost Hunting in Prison: Contemplating Death through Sites of Incarceration and the Comodification of the Penal Past," in J. Z. Wilson, S. Hodgkinson, J. Piché, and K. Walby, eds, *The Palgrave Handbook of Prison Tourism* (London: Palgrave Macmillan, forthcoming).
35 One of the authors had the chance to attend one of these; visitors were a mix of interested professionals and retired members of the public. It is questionable whether the motives of the public visitors were philanthropic or voyeuristic, although possibly a mixture of both.
36 C. Minogue, "Human Rights."
37 Wilson et al., "In Praise of the Carceral Tour."
38 A. Barton and A. Brown, "Dark Tourism and the Modern Prison," *Prison Service Journal* 199 (2012): 44–9.
39 See, for example, Strange and Kempa, "Shades of Dark Tourism."
40 Wilson, *Prison*.
41 M. Welch, "Penal Tourism and a Tale of Four Cities: Reflecting on the Museum Effect in London, Sydney, Melbourne, and Buenos Aires," *Criminology and Criminal Justice* 13, no. 5 (2013): 479–505.
42 Ibid., 484.
43 Brown, *The Culture of Punishment*; Walby and Piché, "The Polysemy of Punishment Memorialization."
44 J. Turner and K. Peters, "Doing Time Travel: Performing Past and Present at the Prison Museum," in *Historical Geographies of Prisons: Unlocking the Usable Carceral Past*, ed. K. Morin and D. Moran (London: Routledge, 2015), 71–88.
45 Ibid., 85.
46 Phillip R. Stone, "It's a Bloody Guide: Fun, Fear and a Lighter Side of Dark Tourism at the Dungeon Visitor Attractions, UK," in *The Darker Side of Travel: The Theory and Practice of Dark Tourism*, ed. R. Sharpley and P. R. Stone (Bristol: Channel View Publications, 2009), 167–85.
47 Ibid.
48 Foucault, *Discipline and Punish*.
49 Brown, *The Culture of Punishment*.
50 See also W. Morrison, "A Reflected Gaze of Humanity: Cultural Criminology and Images of Genocide," in *Framing Crime: Cultural Criminology and the Image*, ed. K.J. Hayward and M. Presdee (London: Routledge, 2010) 189–208.
51 Stone, "A Dark Tourism Spectrum."
52 This was mainly to meet the demand of the larger criminal population, as the number of prisoners in England almost doubled during this time. See R. McGowen, "The Well-Ordered Prison: England, 1780–1865," in *The Oxford History of the Prison: The Practice of Punishment in Western Society*, ed. N. Morris and D. J. Rothman (Oxford: Oxford University Press, 1995), 71–99.
53 We were also interested in the growing trade in ghost tourism within this and other penal tourism sites, which is discussed in more detail in Hodgkinson and Urquhart, "Ghost Hunting in Prison."

54 K. Soothill, "Prison Histories and Competing Audiences, 1776–1996," in *Handbook on Prisons*, ed. Y. Jewkes (Cullompton: Willan, 2007), 27–48.
55 J. Howard, *The State of the Prisons in England and Wales*, 3rd ed. (London: Gale ECCO, 2010).
56 H. Johnston, "Moral Guardians? Prison Officers, Prison Practice and Ambiguity in the Nineteenth Century," in *Punishment and Control in Historical Perspective*, ed. H. Johnston (Basingstoke: Palgrave, 2008), 79–94.
57 The execution of William Saville in 1844 for the murder of his wife and children drew crowds of more than 2000 people, and 11 people were subsequently crushed to death when the crowd was startled. This case had a key part to play in the campaign against executions being held as a public spectacle.
58 H. Johnston, "Introduction: Histories of Punishment and Control," in Johnston, ed., *Punishment and Control in Historical Perspective*, 6.
59 McGowen, "The Well-ordered Prison."
60 R. Sharpley, 'Shedding Light on Dark Tourism," 9.
61 Visitors are given a prisoner number when they start the tour and have to match this up to a real-life prisoner from both the Georgian and Victorian eras. The costumed interpreters will then role-play and quiz the visitors about their crime, and what punishments await them.
62 M. Welch, "Penal Tourism and a Tale of Four Cities."
63 Wilson, *Prison*; Brown, *The Culture of Punishment*.
64 Stone, "A Dark Tourism Spectrum," "It's a Bloody Guide."
65 Barton and Brown, "Dark Tourism and the Modern Prison."
66 Turner and Peters, "Doing Time Travel."
67 Barton and Brown, "Dark Tourism and the Modern Prison": 45.
68 Hodgkinson and Urquhart, "Ghost Hunting in Prison."
69 Stone, "It's a Bloody Guide", 169.
70 Interestingly, the gaol has taken on various guises over the years: for a period of the twentieth century it contained a shop, and was once also a nightclub.
71 E. Lord, *Derby Past* (Chichester: Phillimore, 1996).
72 www.derbygaol.com/crime_punishment.htm (accessed March 7, 2016).
73 Similarly, the Galleries of Justice also relate how the bodies of executed murderers were immediately bought back into the courtroom to be publicly dissected.
74 N. Llewellyn, *Art of Death: Visual Culture in the English Death Ritual c.1500–c.1800* (London: Reaktion, 1997).
75 Foucault, *Discipline and Punish*.
76 Llewellyn, *Art of Death*.
77 Gatrell, *The Hanging Tree*.
78 Ibid.
79 Ibid., 313.
80 Stone, "A Dark Tourism Spectrum."
81 Lennon, "Dark Tourism and Sites of Crime."
82 Lennon and Foley, *Dark Tourism*.
83 Brown, *The Culture of Punishment*.
84 Walby and Piché, "The Polysemy of Punishment Memorialization."
85 Hodgkinson and Urquhart, "Ghost Hunting in Prison."
86 Walby and Piché, "The Polysemy of Punishment Memorialization."
87 Brown, *The Culture of Punishment*.

4 Patrimony, engineered remembrance and ancestral vampires

Appraising thanatouristic resources in Ireland and Sicily

Tony Seaton

Introduction

Much discussion has been devoted to defining thanatourism and examining its characteristic forms (battlefield tourism, holocaust tourism, 'dark' museums, murder and atrocity sites, etc). However, less attention has been focused on thanatourism as a destination image element, in the sense that certain places are partly or wholly branded and known for their association with particular touristic activities, interests and tastes (e.g. golfing locations, surfing and diving resorts, heritage cities, etc). This chapter explores what identity a dark tourism destination might be in perceptual terms through conceptualising it as a form of 'patrimony'. It offers a perspective on the historic resources that might constitute this patrimony, and an approach to inventorying them comparatively, using a simple diagnostic model. The approach is exemplified in a comparison between Ireland and Sicily, which includes a revisionist perspective on one of Ireland's best-known 'dark' resources: *Dracula* and its author, Bram Stoker.

The starting point for this chapter is an axiom that thanatourism is part of the *patrimony* of place. Patrimony means 'inheritance' and is almost a synonym for the English word 'heritage', but is here preferred to the latter for its more affective and holistic associations connoting a strong sense of history, a proprietary commitment to one's past as *space* and *behaviour* and a clear idea of the preferred face of that past and how it should be promoted to the world. Patrimony comprises the tangible and intangible features of place (nation or region) derived from its physical features (scenery, fauna and flora, built environments), its cultural products (literary, artistic, religious, scientific, performative, etc) and the way of life of its people.

Thanatouristic patrimony is a subset of overall patrimony that may be described as the collective annals of *fatality* and *mortality* relating to a place. It comprises 'dark' sites and events remembered or commemorated in narratives and representations delivered in oral traditions, archives, histories, cultural myths and all forms of mediated communication. It includes the sites and events themselves and the biographies of the dead actors who participated in them, as well as the material forms of their public commemoration in churchyards, in cemeteries, on blue plaques and memorials and by signage and street naming. Thanatouristic patrimony can be seen as part of what Horne has explored as 'public culture'.[1] It is the induced remembrance of a society's or community's dead – some of whom will be commemorated for their voluntary acts and others as unwilling victims of fatalities produced by circumstances beyond their control, for example, natural disasters, wars, accidents and other kinds of fatal misfortune. Most, however – the 'Great Majority', as the Victorians labelled the dead – will be forgotten, unless posterity turns a spotlight on their lives that brings them to life in narratives of the present. Tourism narratives are a prime source of these.

Thanatouristic patrimony is unequally distributed among destinations. It is affected by a number of factors, particularly time. Other things being equal, older destinations have greater thanatouristic equity than newer ones, because mortality and fatality in them go back longer. Size also matters, since the more populous a destination, the greater the volumes of mortality and fatality. But there are exceptions to this. Small regions or states lying close to bigger, more politically powerful ones, or on strategically important routes, experience higher levels of recurrent fatality because of their susceptibility to invasion and occupation and the internal divisions these create. Ireland and Sicily are both examples of this 'small, vulnerable neighbour' syndrome, which has made them both 'punch above their weight' in thanatouristic patrimony – or, more exactly, *'have been punched* above their weight' – by their susceptibility to invasion and colonisation. Such destinations suffer and commemorate more intensely felt oppressions than less strategically vulnerable places.

Not all mortality and fatality become patrimony. Historical events selected for remembrance must be orchestrated, through narratives and representations, by those desiring to commemorate them. This engineering of remembrance may be the work of individuals, interest groups or governmental organisations. The most common forms are personal acts of grief and respect by individuals commemorating loved ones and significant others in private funerary and memorial practices. At the other extreme, thanatouristic patrimony may be engineered by institutional agencies seeking to perpetuate the memory of exemplary others as public models, through biographical narratives, place-naming (Wellington, Stalingrad, Victoria Station, Kennedy Airport) and other forms of commemorative promotion which may be exposed to international, as well as domestic, audiences. The choice and number of events and people nominated for remembrance, and the permissible volume of representations and narratives told about them, are always controlled by 'backstage' agents through decision processes that their audiences rarely recognise or reflect on. Thanatouristic patrimony is not static. The choice of remembered subjects and the nature of remembrance may change. Once excluded and proscribed subjects may later be orchestrated as public culture. Until the 1990s the French were allowed few memorials to their dead at the battle of Waterloo, and the Kwa-Zulus were allowed none at all at colonial battlefields in South Africa until Mandela's release.[2] As political and ideological climates change, disclosure of the previously unknown may become high-impact patrimony.

The final factor influencing thanatouristic patrimony is the ratio of historic *mortality* to *fatality*. Mortality is the cumulative total of all deaths in a place, most of which will be from 'natural causes'. Fatality is the total of deaths from sudden, often violent causes. In every place the mortality rate is ultimately 100 per cent since all must die. The fatality rate is normally much lower, but plays a greater part in thanatourism patrimony than the mortality rate. This is because death from natural causes is never of interest to more than a few – the deceased's relatives and friends – except in the case of the few individuals who have found fame and fortune in the public eye.

Despite the efforts of those who engineer them, narratives and representations of remembrance may not achieve the effects intended. Like all forms of heritage they may be misunderstood or contested by audiences, including tourists.[3] Patrimonic narratives may be weakened, negated or created by the superior representational power of competing narratives from elsewhere. A destination's historical identity is always, in part, affected by 'othering' processes, constructed in the political and tourism narratives told about it abroad, as we shall see in the case of *Dracula* and the vampires of Romania.[4] In summary, thanatouristic patrimony may be viewed as the engineered and orchestrated remembrance of fatality and mortality relating to place, commemorated in representations and narratives. It may be promoted by the efforts of private individuals, groups or powerful public agencies.

Inventorying thanatouristic patrimony: the seven domains of Mortality and Fatality

How can this theoretical overview be applied in the comparative analysis of thanatourism at a destination? One way of inventorying existing and potential thanatourism resources is through a diagnostic approach that divides the history of a place into seven *sectoral domains* for exploration, with the aim of identifying significant sites, events and figures that might be represented and narrated as 'dark' patrimony. All the domains but one have been adapted from the well-known PEST model, used in managerial analysis to identify and audit the different environments (economic, political, social, environmental, etc.) that business must take into account in strategic planning.[5] The framework deploys these not as environments, but as differentiated resources of historical mortality and fatality that may be commemorated and narrated as thanatouristic themes. They comprise the geo-physical domain; the geo-social domain; the political and judicial domain; the socio-economic domain; the cultural domain; the religious and ethnic domain; and the funerary domain.

The *geo-physical domain* comprises 'acts of God', fatalities caused by inanimate nature (earthquakes, volcanoes, floods, hurricanes, etc), which vary in scale and frequency by geography and climate. The biggest ones in history are well known, having been engineered for remembrance on an international scale (for example, the destruction of Pompeii and Herculaneum in AD 62 and 79 and the Lisbon Earthquake of 1757). Smaller ones may have more limited currency but become part of the folk memory of the places where they occur, for example Hurricane Katrina in New Orleans in August 2000.

The *geo-social domain* comprises accidental deaths caused by human agency. All destinations occasionally suffer man-made fatalities, but they are more frequent in developed, industrialised countries that combine advanced technology with high urban population densities. These regularly produce disastrous accidents less common in pre-industrial settings and include transport deaths by air, road, rail and sea; fatalities in factory accidents and building collapses; and, in the twentieth century, casualties from chemical, biological and nuclear leaks, explosions and spills. In addition, the pressure of urban numbers may on occasions produce crowd tragedies in public places and spaces (theatres, sports arenas, religious gatherings). Most of these human accidents hardly existed in pre-industrial societies and some of them became thanatouristic *causes celebres* (the Titanic in 1912, the collapse of the railway bridge over the River Tay in Scotland in 1877 and the nuclear disaster at Chernobyl in 1986).

The politico-legal domain comprises fatality and mortality deliberately inflicted on humans by other humans. The most recurrent of these have been wars, which are almost always later commemorated as thanatourism patrimony by their protagonists. War fatalities were once largely confined to battlefield engagements between opposing armies, though massacre, rape and pillage of civilians might follow military victory. In the twentieth century armies ceased to be the main casualties. Civilian populations became targets from land, sea and air, creating new, patrimonic nightmares of remembrance (the London Blitz, the fire-bombing of Dresden, the A-bombing of Nagasaki and Hiroshima). Civil wars and conflicts within states between religious, political and ethnic groups created other kinds of violent patrimony – pogroms, ethnic cleansing and revolutions, which have all had their memorials and ritual commemorations. In addition to the politico-judicial fatalities of war created by external enemies, the legal institutions and punitive systems of states may produce varying volumes of fatality from within, supported by the efforts of violent criminals and those charged with executing violent retribution. Crime and punishment have proved popular tourism narratives in many places, including Alcatraz Gaol tours in San Francisco, visits to the Chamber of Horrors in Madame Tussauds and the bloody history of the Tower of London.

The *religious domain* reflects the sad truth that, despite their nominally pacific and spiritual goals, religions have historically divided populations and created fatalities and mass killing fields. These have included martyrdoms, collective persecutions and ethnic cleansing. Their legacies have been public memorials, exhibited relics and pilgrimages to sacred spaces, all of which may be seen as thanatouristic encounters with mortality and fatality. Jewish Holocaust memorials have multiplied in Poland, Germany and other countries of German occupation, as well as in museums, including one in the main Smithsonian Mall quarter in Washington. One unanticipated consequence of religious thanatourism is that the persecution of heresies perpetuated by a society may become part of its leisure tourism patrimony later. In the past two decades 'Cathar Country' has been developed as a form of destination branding in Southern France in the region where the Cathars, a heretical religious group, were brutally suppressed by the Catholic Church in the Middle Ages.

The *socio-economic domain* of thanatourism derives from the fact that most societal and social systems produce inequities in the distribution of social and economic benefits which intermittently erupt into crises with traumatic, sometimes fatal effects that may enter the annals of collective memory. The Great Depression of the 1930s in America produced suffering and suicides. State-created mass famines in Russia and China went on largely behind the backs of the world in the twentieth century. Such crises may be commemorated for posterity in museum narratives, exhibitions and educational tours. In the past decade socio-economic deprivation has provided a new addition to tourism discourse in the phenomenon known as 'slum tourism', visits by the rich to the homes of the poor. Like thanatourism, it is based on the fascination of the 'other', whether for charitable or voyeuristic reasons – an issue that has been vigorously debated.[6]

The *cultural domain* is a diverse one, comprising representations of death and fatality in the arts, literature, broadcasting and media, and also the mortality of performers who produce, narrate and embody them. Novels, poetry, films and theatre may provide subjects of mortality and fatality that may be engineered as place narratives. The place may be the real setting of a fictional event, such as the Reichenbach Falls, where Arthur Conan Doyle set the apparent death of Sherlock Holmes, but it could equally be a real place associated with the mortality of a real cultural figure, for example the site where a famous author, musician or other celebrity died.

The *funerary domain* is not derived, like the others, from the PEST model, but is a discrete sphere of societal praxis which has an intrinsic relationship to thanatourism, since disposal of the dead is a necessity in all societies which has generated a diverse range of funerary and post-funerary practices at different times and in different cultures. Such practices have included differences in the rituals of burial and body disposal; in commemoration and memorials; and in the length and forms of mourning and remembrance. These funerary variations have fascinated travellers and featured regularly in travel texts. In Europe historic cemeteries are increasingly entering the leisure and tourism economy as part of the patrimony of place, a heritage role that was promoted by the establishment of the Association of Significant Cemeteries in Europe in the 1990s, which has over 300 subscribing cemeteries as members.[7]

In summary, domain appraisal offers a diagnostic approach for destination planners in identifying categories of 'dark' events, sites and biographies that may be engineered as place narratives and representations. The categories are provisional and may be modified or revised. Some may overlap and be hard to distinguish from each other; examples include the political, religious and socio-economic domains. Nevertheless they provide a useful starting point for identifying and auditing 'dark' resources. It is unlikely in major destinations that domain analysis will often disclose narratives of mortality/fatality that have previously escaped notice. It may, however, highlight weaknesses, or new opportunities, in the exploitation of existing ones. The success of the *Titanic* exhibition, opened in 2012 at the Harland and Wolf Shipyard in Belfast, where the ship was built, to commemorate the 100th anniversary of its sinking is a good example of how a link with the doomed liner that had always been known about could in time be developed as a mega-attraction. The case of *Dracula* in Ireland, to be discussed later, could be a latent patrimonic resource still to be recognised and developed. In smaller destinations (counties, cities, towns), historical research may reveal new and unusual subjects that can be engineered as 'dark' narratives. In addition, there are instances where existing narratives for certain kinds of attractions, not previously positioned as 'dark' sites (e.g. heritage, scientific and literary attractions), may widen their appeal through 'dark' thematisation. The National Trust, one of the biggest heritage organisations in Britain, has made tentative steps toward advertising some of its many historic properties using more populist 'dark' narratives.[8]

In applying domain analysis the starting point is to employ or engage researchers to recognise and inventory historical events, sites and figures that may be adapted as 'dark' destination narratives and representations. They may be tourism officers, heritage consultants, local historians or academic specialists with experience of trawling diverse historical materials for promising themes. These will include old guidebook archives, local histories and special library collections that include historic travel memoirs, geographies, and so on.

Domain analysis: Ireland and Italy

Comparative domain analysis may be exemplified in illustrations from two island destinations, Ireland and Sicily, which display interesting contrasts and similarities. Both have shared a history of colonial oppression, urban and rural poverty, domination by aristocratic and commercial elites, internal conflicts and periodic natural catastrophes. In geo-physical terms Sicily is a more savage terrain than Ireland, with climatic extremes of heat and cold, and periodical catastrophic fatalities from earthquakes, floods and volcanic eruptions. The first major English guidebook to Sicily included a twenty-page history of Etna eruptions from the sixth century BC to 1852.[9] The island also acquired an image for criminality as a land of brigands in the nineteenth century. As a result the island became something of a 'dark' attraction in the early twentieth century for rich tourists from Britain and America who had already seen the conventional 'sights' of mainland Italy. Douglas Sladen, the leading author of guidebooks to Sicily in Edwardian times, described ironically how these prototypical 'dark tourists' were more pumped up by the edginess of Sicily than by glossy advertising for its conventional pleasures: 'The odd thing is that, while no amount of advertising hotels, or hot springs, or sports clubs, seems to do any good, floods, earthquakes and brigands have their value.'[10]

This semi-flippant observation was written after minor earthquakes and floods in 1903. Five years later a catastrophic undersea volcanic eruption changed the scale of disaster and produced a tsunami that devastated the historic Sicilian city of Messina, killing half of the city's 200,000 inhabitants. It made international headlines and drew tourists, volunteers and support groups, especially from the USA. Ireland had no comparable geo-physical disaster with such immediate impact, but its 'potato famine' in the 1840s produced greater numbers of fatalities through starvation and disease, as well as causing around 25 per cent of its population to emigrate. It also attracted tourists and charitable groups, including Quakers who risked their own lives working in epidemic conditions among the sick.

Imperial oppression in Ireland and Sicily produced 'dark' events that became part of the domain of the politico-legal patrimony of both. In Lampedusa's Sicilian novel *The Leopard*, one of his characters deplores 'the thousand invasions we have had' – an exaggeration, perhaps, but one founded on the history of the three millennia during which Sicily was successively occupied by Phoenicians, Carthaginians, Greeks, Romans, Arabs, Normans, French, Spanish, Germans and English. Ireland had fewer unwelcome guests, but those they had – the Romans, the Vikings, the Normans and the English – left legacies of poverty for the natives and wealth for colonising elites, as well as provoking rebellion and colonial suppression, with reprisals on both sides. The battles, massacres and intermittent rebellions created by colonial conflicts became tourist narratives in the important Victorian guidebooks of John Murray, Baedeker and A&C Black. Murray's first Sicilian guidebook, written by George Dennis after sixteen years of research and published in 1865, was a huge work, over 500 double-column pages long. It included an account of the 'Sicilian Vespers', a massacre in Palermo of occupying French forces which took place on Easter Tuesday 1282 at an outdoor gathering in which the native population turned on their French occupiers and slaughtered 3,000 of them.[11] The first 'picturesque' guidebook to Ireland, published in the long-running series by the Scottish publisher A&C Black in 1855, recorded the massacre at Drogheda in 1649 when the city was stormed by 12,000 of Cromwell's troops, with 2,000 Irish deaths.[12] These are only two historical instances of the politico-legal domain as 'dark' political touristic narratives. A modern example of how 'dark' politics may quickly become patrimony is the aftermath of the murder in 1992 of two Sicilian anti-corruption judges, Borsalino and Falcone, by the Mafia. The deaths of these two brave men created such an outcry that the national airport at Palermo was renamed in their honour, where it remains a memorial for all Sicilians, as well as for tourists.

One of the most distinctive features of Ireland and Sicily for visitors, compared to many other European destinations, has always been in the funerary domain. Sladen described Sicily as 'the land of tombs and tombless corpses', and in his encyclopaedic guidebook catalogued the island's patrimony of catacombs, cave sepulchres, memorials and campo santos in detail, region by region.[13] They included the catacombs of Syracuse, which were larger than those at Rome, and the grotesque mummified cadavers in the Convento dei Cappuccini, Palermo, where tourists could view in an underground crypt the 'most perfect collection of bodies dried in Cappuccine fashion'.[14] Another traditional Italian and Sicilian funerary institution visible to tourists was the Burial Guild, which survived for centuries. This was a local society, or 'confraternity', formed by its members to save money to pay for family funerals. When a funeral took place the members bore, or followed, the coffin of the deceased through the streets, wearing white, hooded dresses covering everything except their boots, eyes and mouths.[15] It looked disconcertingly like a Klu Klux Klan procession and illustrations of it are common in guidebooks going back to the eighteenth century.

Ireland was equally well known among tourists of the past for its funerary traditions and customs, which included carnivalesque processions, 'professional' wailers and bibulous wakes for the dead. More visibly, Ireland included a great number and variety of funerary sites, old and new. Murray's guide described them at length for Victorian tourists, dividing them into two categories: 'Sepulchral', comprising 'Cromlechs, Caves, Mounds and Cairns', and 'Memorials', which included 'Pillars, Steles, and Inscribed Stones'.[16] Glasnevin was the first modern cemetery to be included in a major tourism guide, in 1855.[17]

The funerary domain was closely associated in Ireland and Sicily with the religious domain. Death, as much as life, was shaped and policed by an authoritarian brand of ultra-Catholicism under the control of the Pope in Rome, through a network of priests and monastic orders throughout both lands. Ireland became a by-word for piety with its monastic traditions, Christian saints and martyrs, sacred sites and 900 holy wells.[18] One of the main de facto tourism promoters was the Catholic Truth Society, operating from Dublin, which published a series of cheap illustrated pamphlets in the early twentieth century, each of which promoted different religious sites. They were printed in their thousands.

In Sicily the Catholic Church exerted the same kind of influence through the same religious networks of priests, churches and monastic orders. In addition to this ideological dominance, the church had its main coercive institution, the Inquisition, based in Palermo, where it conducted trials, torture and executions until its abolition in 1782. The cultural domain in Ireland and Sicily encompasses the ways in which both islands have been represented in narratives and representations touching upon death. The field is enormous, particularly for Sicily, where instances extend back to classical antiquity and the work of Homer, Virgil and Ovid, which included poetic accounts of the eruption of Etna and the myth of Scylla and Charybdis, two rocks north of Messina around which, according to legend, violent whirling currents threatened sailors at sea. They extend forward to gangster films from Hollywood and reports of Mafia trials in the press and TV (the latter being examples of domain ambiguity, in that they could be classified as politico-legal or cultural in type).

In Ireland, ancient myth and folklore included violent narratives of legendary heroes and villains which were periodically recycled by later writers, including Yeats. Less well known are the 'dark' cultural links Sicily and Ireland shared through their inadvertent participation in the development of the Gothic novel, which, from its beginnings between 1760 and the 1830s, later evolved into horror, ghost and crime fiction. Early Gothic novels typically featured innocent young virgins threatened by corrupt monks, decadent aristocrats, midnight hauntings, murder and financial ruin. They have partly been seen as Protestant propaganda, with many set in southern Europe, especially Italy and Spain.[19] Sicily featured in the title of several, including *The Sicilian Boy*, *Sicilian Pirate* and *Sicilian Romance*.[20] More than a century later Giuseppe Lampedusa produced a less sensational, modern version of Sicilian Gothic in *The Leopard*, which traced the downfall of a great aristocratic dynasty. It led one critic to write: 'Death is the first sentence of the book and permeates the rest of the novel.'[21]

Ireland's contribution was less the setting of Gothic texts than the fact that it was the homeland of three of the most important Gothic authors – Charles Maturin, Sheridan le Fanu and Bram Stoker – a trio who thus represent a powerful presence in 'dark' cultural patrimony. It was fitting that in 2009 the *Irish Journal of Gothic and Horror Studies* was launched as the first academic publication celebrating the Gothic in a nationalistic context, though its editorial remit is by no means limited to Ireland. All of these domain features have helped to give Sicily and Ireland an 'otherness' in the eyes of travellers; a difference from more 'developed' countries in Europe. 'One of the great charms of Sicily is its un-Europeanness', wrote Sladen in 1905.[22] It was this exoticism that perhaps led notorious self-styled satanist Aleister Crowley, who called himself the 'Great Beast', to take his 'princess' and his occult practices in the 1920s to Cefalu, near which he set up his 'Mystic Abbey of Thelema' in a nondescript house just below the level of the road running outside. He remained there until being deported from Italy as an undesirable alien in May 1923, but survives today as an unusual contributor to the dark materials of Sicily's thanatourism patrimony.[23]

Dracula and Ireland's Gothic patrimony

Dracula first appeared as a novel by the Irish writer Bram Stoker, and might well now be the best-known feature in Ireland's 'dark' cultural patrimony. On publication in 1897 its success was immediate,[24] and has since extended to theatre, radio and more than eighty film versions that have made the name of its eponymous villain as well known internationally as James Bond.[25] The book has never been out of print and now attracts increasing academic attention, as well as retaining its popular following. Yet it has hardly featured as part of Ireland's literary patrimony in the way that the Brontes, the Bloomsbury Group or Thomas Hardy have become tourism fixtures in England. In Ireland tourism narratives associated with *Dracula* have been confined to biographical details of Stoker's life in Dublin – which he left when he was 35 – rather than his works. International audiences know little of Stoker, not even that he was Irish.[26] Speculations about the cultural origins of Dracula as a character have achieved currency have mainly focused on eastern Europe and Vlad the Impaler, rather than origins in Ireland.[27] This has led the literary critic Declan Kiberd to observe that 'the scale and scope of the *Dracula* myth have been so enormous as to obscure not only the author but the Irish world that made him'.[28]

The failure to associate Stoker and his book with Ireland has allowed two destinations – one in England, the other in Romania – to establish touristic ownership of *Dracula* as Gothic patrimony. Whitby, a northern seaside resort in Yorkshire, has in recent years branded itself as the Goth and vampire capital of England, on the basis it that it was the port to which the vampire Count Dracula sailed in his fictitious pursuit of his virginal victims. The resort stages an annual week-long 'Goth festival' as part of its vampiric celebrations. The other destination which has, almost reluctantly, come to be seen as the authentic location of vampire lore and legend is Transylvania in Romania. This has happened less because Romanians have engineered it than through representations orchestrated in the West.[29]

Why this failure by Ireland to secure ownership of one of its most famous literary properties? The main reason may be that until quite recently *Dracula* was considered as 'pulp fiction' rather than 'literature', and received little serious critical attention, compared with the industrial scale of literary, academic and touristic investment in Wilde, Joyce and Yeats. It is only since the 1970s that academics have embraced popular culture as an important field of study. The other reason is that, though Stoker was born and bred in Dublin, there was nothing in *Dracula* that was seen as distinctively Irish – despite the fact that the Gothic bibliophile Montague Summers, in his study of European vampirism, expressed the opinion that in Ireland the vampire had once been known as Dearg-Dul ('red-blood sucker'), and his ravages had been universally feared.[30] However, the cultural roots of *Dracula* have recently come under increasing scrutiny as a result of website discussion and blogs, with results that may re-order its place in Irish literature and 'dark tourism' patrimony. Literature is typically engineered as a touristic resource in two ways. One is by promoting the locations of the author's life, a procedure that has already started with limited success in relation to Stoker's Dublin addresses. The other is by making the locations of the literary works the focus. In Victorian times, travelogues using one or both of these strategies were known as 'homes and haunts' studies, and many writers were the subject of them.[31]

Over the past two decades academic research has implicitly addressed the question of *Dracula's* Irish antecedents through engaging with the history of vampire lore and legend in medieval Ireland. In 2003 Peter Carey surveyed more than a dozen different academic papers on lycanthropy, vampires and werewolves in Irish history, including a classic one by Reinard, first printed in 1936.[32] His results suggested that legends of lyncanthropic behaviour in secular and religious texts date back almost 1,000 years. The best-known of the early sources Carey cited was a travelogue by Gerald of Wales, who in 1188 wrote one of the earliest accounts of Ireland.[33] In it is included the strange story told to him by a priest of an encounter with a 'wolf man' who addressed him in a human voice, saying he belonged to a group of men in Ossory who every seven years had to abandon human shape and become wolves. Carey juxtaposed this with a Norse work written about the same time, which told the story of St Patricius, who, while preaching Christianity in Ireland, had been confronted by men howling like wolves at him as he passed by.[34] A famous English source cited by Carey was Camden's *Britannia*, the first great county-by-county atlas and gazetteer of England, Wales, Scotland and Ireland, printed and reprinted in the late sixteenth and early seventeenth century. Camden was a schoolmaster who pursued historical research during his vacations, travelling around Britain gathering information on the past from regional witnesses. His results included evidence that lycanthropy was held in serious regard in Ireland. Though he was sceptical himself, he duly reported what he had heard:

> Whereas some of the Irish, and such as would be thought worthy of credit, doe affirme, that certyaine men in this tract are yeerely turned into Wolves: surely I suppose it to be a mere fable: unless haply through that malicious humour of predominant unkind Melancholy, they be possessed with the

malady that Physicians call Lycanthropia, which raiseth and engendereth such like phantasies, as that they imagine themselves to be transformed into Wolves. Neither dare I otherwise affirme of those metamorphosed Lycaones in Liveland, concerning whom many Writers deliver many and marvellous reports.[35]

Carey suggests that Camden's information may have come indirectly from Gerald of Wales, but regardless of where it originated, the reference extends the textual time-span of Irish lyncanthropic myth by another three centuries after Gerald of Wales, and the reference to the 'unkind Melancholy' of Lycanthropia anticipates the characterisation of Dracula as depicted by Stoker. His melancholic isolation and inability to control or end his vampiric compulsions lends him a psychological credibility, previously assumed to be Stoker's invention, but one which may have also derived from traditional vampiric myth in Ireland.

Another source on vampiric precedents in Ireland comes from an older work not included in Carey's review. This is the *First Synod of St Patrick*, a Latin manuscript going back to AD 457, which has been called the 'earliest surviving document concerning ecclesiastical discipline in Ireland'.[36] It is a sort of crime and punishment handbook for churchmen, setting out behaviour in which they were forbidden to engage, and the punishments to be meted out if they did. The manuscript resembles 'penitential' texts that were circulated as rule-books for clerical communities. It included thirty-four different 'don'ts', the longest of which was about vampires:

> A Christian who believes there is such a thing in the world as a vampire, that is to say a witch, is to be anathematized – anyone who puts a living soul under such a reputation; and he must not be received again into the Church before he has undone by his own word the crime that he has committed, and so does penance with diligence.[37]

The First Synod was initiated by two bishops, Auxilius and Iserninus, at the time of St Patrick's Irish Mission, and is believed to have had his 'express approval'.[38] It thus links vampires to Ireland's patron saint more than 500 years before any of the citations discussed above. Another strand of support for the Irish influence on Dracula is that of Ginzberg, who concluded from his study into vampires and lycanthropy in 1997 that there was a 'massive presence in Ireland of legends linked to werewolves'.[39]

The Irish evidence has its parallels with findings published by English academics from the universities of Oxford and London which were featured in an hour-long TV documentary, 'Britain's Medieval Vampires', which was broadcast on Halloween night in 2015. The programme linked Reynolds' work on 'deviant burial customs' in Anglo-Saxon graves with Moore's work on fear of the dead arising. The thesis was that evidence of bodies mutilated or placed in unusual positions within graves was evidence of the need to make sure they did not return to earth, thus suggesting a belief in vampires. Montague Summers had commented on pagan graves in Ireland buried under stone cairns intended to entomb corpses securely, which reflected a

Figure 4.1 'Natural history of two species of Irish vampires'. Ireland oppressed by two vampires – English landlordism, and the Church and Law. Satirical cartoon published in *The Looking Glass*, 1831. Author's collection.

similar fear of vampiric reappearances.[40] A rare visual allusion to vampires in Ireland also appeared in an English satirical magazine, 'The Looking Glass', in 1831. It included a cartoon with the title 'Natural history of two species of Irish vampires' and showed two grotesque vampiric figures, identified as a landlord's agent and a 'law priest . . . from a neighbouring Country', feasting on and sucking the blood of 'Erin', a maiden, who represented Ireland.[41] Nearby a sign reads: 'Oppression Brought Famine and Pestilence.' See Figure 4.1.

The antecedents of *Dracula* may thus partly lie in Irish literary and legendary influences going back 1,500 years, connecting its cultural roots less to Vlad the Impaler's Romania and Whitby in Yorkshire and more to native traditions. These historic resources could, with the cooperation of cultural historians, Irish literary specialists and tourism planners, be developed as new narratives around a Gothic Ireland concept, not just based on the life and work of Stoker, but also drawing on materials by Maturin and Le Fanu. This literary 'Gothic Ireland' theme could also be extended into the visual arts by including Ireland's very considerable Gothic architecture patrimony, as well as including some of the other 'dark' domains discussed in this chapter. Promotional support to articulate and disseminate the overall destination positioning proposed would include a *Guide and Directory to Gothic Ireland,* and perhaps sponsorship of a Gothic writer and artist in residence.

Conclusion

This chapter has offered a perspective on thanatourism as a holistic destination concept and a framework for appraising the 'dark' potential of destinations through the concept of 'thantouristic patrimony', conceptualised as the engineered remembrance of fatality and mortality. Thanatouristic patrimony can be identified and audited by analysis of seven historical domains in order to reveal 'dark' materials for place narratives and representations. Historical features in Ireland and Sicily were compared as practical illustrations of the approach. In Ireland, an examination of recent work on the history of vampire myths disclosed evidence that may inject new blood into the search for the native origins of one of Ireland's best-known cultural exports. If the search is successful, the lycanthropic undead may be about to reappear in Irish place narratives after an uncanny exile in Whitby and Romania.

Notes

1 D. Horne, *The Public Culture: The Triumph of Industrialism* (London: Pluto, 1986).
2 A. V. Seaton, "War and Thanatourism: The Waterloo War 1815–1914," *Annals of Tourism Research* 26, no. 1 (1998): 29; A. V. Seaton and J. J. Lennon, "Thanatourism in the Early 21st Century: Moral Panics, Ulterior Motives and Alterior Desires," in *New Horizons in Tourism. Strange Experiences and Stranger Practices*, ed. T. V. Singh (Wallingford: CABI, 2004), 74.
3 G. Ashworth and J. Tunbridge, *Dissonant Heritage: The Management of the Past as a Resource in Conflict* (Chichester: Belhaven, 1996).
4 D. Light, *The Dracula Dilemma: Tourism, Identity and the State in Romania* (Aldershot: Ashgate, 2012).
5 The Patrimony model is adapted from the PEST model of environmental scanning. The seminal formulation of PEST was by Francis Aguilar, in a study that hypothesized four principal "environments" which organizations had to accommodate and manage in their operations: namely the Political, Economic, Social and Technical environments; hence the acronym PEST. Other theorists later extended the PEST model by adding to the number of environments e.g. legal, demographic, intercultural, ethical. Collins has provided a tabular resumé of expanded versions: R. Collins, 2010, *A Graphical Method for Exploring the Business Environment*, accessed November 1, 2015, http://users.ox.ac.uk/~kell0956/docs/PESTLEWeb.pdf
6 See F. Frenzel, K. Koens, and M. Steinbrink, *Slum Tourism: Poverty, Power and Ethics*, (London: Routledge, 2013).
7 T. Seaton, with M. North and G. Gajda, "Last Resting Places? Recreational Spaces? Or Thanatourism Attractions? The Future of Historic Cemeteries in Europe," in *Landscapes of Leisure. Space, Place and Identities*, ed. S. Gammon and S. Elkington (Basingstoke: Palgrave, 2015), 71–95.
8 The National Trust is a leading heritage organisation in England which manages properties across Britain that include country houses, castles and monastic sites. In recent years it has targeted its efforts more broadly to attract the mainstream family market rather than mainly inveterate culture seekers. In 2004 it ran an advertisement featuring illustrations of a castle and country house dominated by the demotic headline: 'Come home with some great holiday stories like Adultery, Murder, and Ghosts . . . You'll take home something more interesting than a stick of rock.'
9 G. Dennis, *A Handbook for Travellers in Sicily* (London: Murray, 1865).
10 D. Sladen and N. Lorimer, *Queer Things about Sicily* (London: Anthony Traherne, 1905), viii.
11 G. Dennis (1865), op. cit., 103–105. For a more modern account see S. Runciman, *The Sicilian Vespers* (Cambridge: Cambridge University Press, 1958).
12 Anon, *Black's Picturesque Tourist in Ireland* (Edinburgh: Adam and Charles Black, 1855).

13. D. Sladen, *Sicily: The New Winter Resort. An Encyclopaedia of Sicily* (London: Methuen, 1905), 58.
14. Ibid., 410.
15. Ibid., 126.
16. Anon, *Handbook for Travellers in Ireland*, 4th ed. (London: John Murray, 1878), 35–6.
17. Anon, *Black's Picturesque Tourist in Ireland* (1855), op. cit., 35–6.
18. For modern accounts see M. Carroll, *Irish Pilgrimage. Holy Wells and Popular Catholic Traditions* (Baltimore: John Hopkins University Press, 1999); P. Harbison, *Pilgrimage in Ireland. The Monuments and the People* (London: Barrie and Jenkins, 1991).
19. See V. Sage, *Horror Fiction in the Protestant Tradition* (New York: St. Martin's Press, 1988), for an extended account of the gothic and Protestantism.
20. M. Summers, *A Gothic Bibliography* (London: Fortune Press, n.d.), 502–3.
21. Dustwrapper review of the 2nd English edition of *The Leopard*, 1962.
22. D. Sladen (1905), op. cit., 27.
23. J. Symond, *The Great Beast: The Life of Aleister Crowley* (New York: Rider and Co., 1951).
24. M. Summer, *The Vampire* (London: Kegan Paul, 1928), 331–7.
25. R. T. McNally and R. Florescu, *In Search of Dracula. A True History of Dracula and Vampire Legends* (London: New English Library, 1973), 214–24.
26. This was the finding of three "quiz night" surveys in England, asking people who wrote *Dracula* and what nationality s/he was.
27. McNally and Florescu, *In Search of Dracula*, 214–24.
28. D. Kiberd, *Irish Classics* (Cambridge, MA: Harvard University Press, 2001), 384.
29. D. Light (2009), op. cit., passim.
30. M. Summers, *The Vampire in Europe* (London: Bracken Books, 1929), 117–22.
31. Writers whose residential locations or literary sttings have been the subject of 'homes and haunts' travelogues include Walter Scott, Alfred Lord Tennyson, Thomas Hardy and Mary Webb.
32. J. Carey, "Werewolves in Medieval Ireland," *Cambrian Medieval Celtic Studies*, Winter (2002): 37–72; J. R. Reinard and V. E. Hull "Bran and Seeolang," *Speculum*, 11 (1936): 42–58.
33. Gerald of Wales, *The History and Topography of Ireland* (London: Penguin, 1982).
34. J. Carey (2002), op. cit., 45.
35. W. Camden, *Britain A Chorographical Description of the most flourishing Kingdomes, England, Scotland and Ireland* (London: George Lowther, 1637), 83.
36. L. Bieler, *The Irish Penitentials, Scriptores Latin Hiberniae* Vol 5 (Dublin: Dublin Institute for Advanced Studies, 1975), 2.
37. Ibid., 57.
38. Ibid., 37.
39. C. Ginzberg, *Ecstasies. Deciphering the Witches Sabbath*, trans. R. Rosenberg, quoted in Carey, "Werewolves," 37.
40. The programme *Britain's Medieval Vampires* went out at 9 p.m. on Saturday October 31, 2015. The studies on which it was based were A. Reynolds, *Anglo-Saxon Deviant Burial Customs* (Oxford: Oxford University Press, 2009) and J. Blair, "The Dangerous Dead in Early Medieval England," in *Early Medieval Studies in Memory of Patrick Wormald*, ed. S. D. Baxter (Farnham: Ashgate, 2009), 539–59.
41. *McLean's Monthly Sheet of Caricatures or THE LOOKING GLASS* 2, no. 20, August 1831.

5 Death camp tourism

Interpretation and management

Gregory J. Ashworth and John E. Tunbridge

Bringing together tourism and the camps

To examine tourism in relation to the heritage sites associated with the Jewish Holocaust of 1933–45 is to bring together two very different phenomena. Even to use the words Holocaust and tourism in the same sentence is necessarily incongruous. The juxtaposition of an entertainment activity with the heritage of human cruelty and suffering seems ostensibly to be bizarre and distasteful. To link the heritage of organised mass murder with an entertainment industry is to introduce a discordant element of seriousness into a fun activity and, more unacceptably, to trivialise the serious memorialisation of horror and tragedy.

Yet this applies to many tourism sites and experiences that have been labelled 'dark', and furthermore the heritage of the Holocaust is inevitably a tourism attraction, either actually or potentially. The sites and artefacts are designated, collected and presented in order to be experienced by visitors, otherwise they would not have been created or continue to exist. It is a heritage that is intended for consumption by visitors in various ways for various objectives. The importance of the topic and the inherent tension and difficulties of its management stem from this juxtaposition and these resulting contradictions, which must first be understood if they are to be managed.

This chapter narrows the focus to one manifestation of the tourism use of Holocaust heritage, namely the Nazi camps, in which the victims were concentrated, accommodated and ultimately largely exterminated. Of course, the relics of the camps are only one aspect of the heritage of this genocide, which would also include the settlements, ghettoes, assembly points and transit routes of the victims, and the subsequent memorials both within and beyond the territories directly involved. Also, the camps accommodated many others as well as Jews, and are thus not exclusively Jewish heritage, although the 'death camps' created after the conquest of Poland (most notoriously Auschwitz) were primarily designed to kill Jews. Having established the topic, the chapter outlines the wider and different contexts of heritage, and of tourism, as a user of such heritage. The next step is to examine the motives for such a heritage tourism among both tourists as the consumers and those creating and managing the sites as producers. This conceptual discussion can then be applied to the sites themselves as geographical locations and physical structures. It becomes clear that the ways in which tourism and heritage have been linked through interpretation are not constant, but have changed and are changing through time, which raises further questions.

The contexts of the camps

As the contemporary use of the past, heritage is a process whereby selections are made from relics, and accounts of the past imagined from the present in order to satisfy present needs. The question immediately arises as to why past human violence, with the resulting human suffering, should be deliberately selected for contemporary memorialisation. Unhappiness and pain have been the lot of most of humanity, most of the time, and death is a universal inevitability. Why, therefore, should people wish to be reminded? Which and whose present needs is such a heritage satisfying? And, as with all heritage, the question of who is selecting, for what reasons and on whose behalf, must always be posed.

Three concepts may be helpful in understanding these contexts and answering these questions. The first is the concept of dissonant heritage, which is a lack of congruence between people and the heritage that they claim or accept as theirs.[1] Such dissonance may occur for various reasons that are relevant here. Heritage associated with violence can be dissonant to many for many reasons, ranging from a strong emotional association with victims or perpetrators to a vague discomfort with the distasteful. The idea of dissonance thus stresses discrepancy, incongruity and discordance, which people will resolve by adjusting their patterns of behaviour to reduce dissonance and increase congruence. The implications of this for management are discussed below.

Second, all violence by and directed towards people is both provoked by and evokes strong emotions, which thus renders it more noticeable and more memorable than most human actions. Violence is, to say the least, unpleasant – most obviously for victims and those who could identify with them, but also for others, such as spectators, who may not be directly involved but nevertheless empathise to varying degrees or experience remorse for their failure to intervene. However, there is an inherent ambivalence in people's attitude towards violence and its consequences. It contains elements of both aversion and attraction: it is simultaneously to be avoided and sought and, as heritage, to be forgotten and remembered.[2] It is this tension that renders violence a powerful if complex element in the management of such heritage.

Third, the idea of atrocity heritage adds three further dimensions to violence, namely magnitude, conscious decision and memorability, which in turn may result from the size or circumstances of the violent event, which is also neither accidental nor incidental. The heritage of genocide is one category of atrocity heritage and adds a further element, namely the intent that all members of a specified group are actual or potential victims, while all members of another group are actual or potential perpetrators. In these cases particularly, bystanders/spectators may feel the magnitude and memorability of the heritage from empathy, shame of inaction or shared experience of historical consequences, all of which are powerfully present in the Jewish genocide.

The notion that an unpleasant, painful and divisive past cannot other than be remembered and memorialised just because it happened is also incorrect, at least in the short term. Thus the questions 'why remember such violent unpleasant events?', and even 'why not forget them?', must be posed. There are many reasons why individuals and groups may wish not to remember and even attempt to forget. Distress, pain and ultimately death are the unavoidable experience of us all, and the lapse or erasure of memories of unpleasant pasts and reminders of unpleasant futures would seem more in the current interests of individuals and their societies. Such heritage is inherently socially divisive, setting victims, and those associating with them, against perpetrators and their associates, and, as noted, variously compromising bystanders. This may mundanely hinder the cohesive functioning of that society. As heritage is always a selection by the present, there are many instances in which the choice to forget rather than remember as a deliberate act of will has been selected by individuals and collective agencies. The avoidance of difficult bits of the past through a policy of individual erasure of memory or collective amnesia, the converse of collective memory, may be justified on many grounds. So too may the 'truth and reconciliation' policy option, as in post-apartheid South Africa, which sought short-term memory to effect long-term closure. Therefore the deliberate choice to remember needs to be overtly justified and the interests of the beneficiaries made explicit.

The context of the Jewish genocide as heritage resource

There have been many genocides in recorded history, but the word was not in common currency before 1948.[3] The genocide of the Jewish people between 1933 and 1945 has a number of characteristics that render it particularly effective as heritage. First, there are visible, visitable, physical remains, notably in the buildings and structures, whether reconstructed in situ or not, of the transit, work and extermination camps. There are also structures and locations associated with the ghettoes in which Jews lived, worked and worshipped for many centuries and to which they were re-consigned prior to extermination. Heritage can of course be created without physical remains and sacralised locations, but their existence makes communication easier. Second, this genocide occurred over a wide spatial extent. It was not confined to a small group in a restricted location but involved in one way or another the entire continent of Europe and, through the diaspora, beyond. It was thus a global phenomenon in a way that many other such events previously and subsequently – including genocides in Armenia, Cambodia or Rwanda, however distressing and locally disruptive – were not. Third, the Jewish genocide has been illuminated by living testimony of both victims and perpetrators, privately and publicly. Fourth, the creation of the state of Israel (1948) provided a national entity that represented the victims, to an extent being legitimated by them.

A further characteristic of this genocide relevant to its interpretation is that the heritage and the people have become dislocated, largely as a result of the genocide itself. Most who would claim this heritage as theirs no longer inhabit the places where it is located, and conversely, those who now inhabit these places may have little claim upon, or identification with, the heritage. This is enclave/exclave heritage, which allows two forms of potential dissonance to exist. First, people, in this case Jews, feel an attachment and make a claim upon heritage in a different country and even continent over which they have no direct control. Second, the people now living predominantly in central Europe occupy places upon which outsiders have lodged a heritage claim. Such dislocations of people and their heritage are not uncommon but the content of the heritage of genocide makes the potential impacts more serious, leading to frustration and anxiety among those making the distant heritage claim and indifference, alienation and even aversion among those in the localities being claimed.

Tourism contexts

Tourism associated with Holocaust attractions can be placed within a number of different tourism contexts. Many 'adjectival' tourisms can be recognised on the basis of the experience of the place-based resource that is commodified into the tourism product. Among the many such tourisms relevant here is travel motivated by curiosity or desire to know ('edutourism'), especially about the past and family links to it ('roots tourism'), or where publicised newsworthy places and events are visited ('current affairs tourism'). Historic sites associated specifically with violence can generate many forms of tourism. These would include 'war tourism' and its localised variant, 'battlefield tourism'; 'disaster tourism', where the attraction is the traumatic human consequences of natural or manmade catastrophes; and even 'danger tourism', where the excitement is the association of violence with personal risk, which possesses some of the characteristics of extreme sports. Closer to the topic here would be 'atrocity tourism' and variants such as 'killing-fields tourism'.[4]

Among those for whom the trip takes on the characteristics of an individual pilgrimage, there are two opposite reactions dependent upon the identification of the visitor. There is 'victim tourism', in which the visitor identifies with the victims – which can become 'grudge' or even 'revenge tourism' when an identified perpetrator is blamed. Such identification with victims of past violence may extend into the present, defining and strengthening the solidarity of the victimised group and even justifying the group's present actions. The converse of this is '*mea culpa* tourism', in which those identifying in some way with the perpetrators of past violence are motivated by a desire for confession, recompense and reconciliation of past violence and its present consequences.

The problem with constructing such a supply-side typology, which could be extended, is that each category of attraction identified may attract quite differently motivated tourists. An attempt to bridge this diversity and bring together a wide variety of sites that shared a common characteristic was the coining of two portmanteau terms, namely 'dark tourism', attributed to Foley and Lennon,[5] and 'thanatourism', attributed to Seaton.[6] Since then an ever-wider range of tourism sites and experiences have been included under these labels, as recently comprehensively reviewed by Hartmann.[7] The question here, however, is whether the visitor experience in the camps sits comfortably under either label, and thus whether this tourism is just one expression of a broader category. Death is an unavoidable but not exclusive element of this tourism product. Thus 'thanatourism' is an accurate but incomplete categorisation. The quality of darkness depends upon the motivations, experiences and especially emotions of individual visitors, before, during and after the visit, and also upon the intentions and interpretations of the producers of the heritage tourism product. This raises the difficulty not only that visitors will experience the site differently, but also that these experiences and evoked emotions may be felt as dark, light or anything in between.[8]

Tourism to the camps may contain elements of many or all of these types of tourism and even be a part of a wider 'Holocaust tourism', including visits to other sites and memorials associated with this genocide. There has been a post-Cold War, post-*Schindler's List* (book 1982, film 1993) boom in such visits, with the major sites registering record numbers (around a million at the Jewish Museum, Berlin and the Anne Frank House, Amsterdam, and around one and a half million at Auschwitz-Birkenau, the most popular of the camps). Despite this and the existence of specialist tour operators, often trading as 'Holocaust tourism', such a tourism is difficult to recognise not only because of the variety of visitor motives but also because many visitors are casual excursionists combining a visit to the camps with quite different tourism elements. Also, although the heritage of the Holocaust is strongly associated with Jewish heritage, not least among visitors, the two are not synonymous. A Jewish 'roots tourism' is both wider and narrower than Holocaust tourism, including other aspects of Jewish life but excluding non-Jewish elements. Thus, despite the variety of perspectives on dark tourism, the fundamental motivational issue remains. It is necessary, therefore, to have some understanding of the motivation, experience and consequent behaviour of the participants in this aspect of Holocaust tourism before it can be labelled as dark.

The motives of visitors

Although our concern here is principally with the collective uses of a communally selected past and its management for collective goals, the importance to the individual should not be overlooked. Violence may be deliberately remembered and memorialised because of its appeal as an individual psychological 'settlement of memory' through mourning, atonement and closure (as typified by cemeteries and by spontaneous memorials at the sites of violent death). However, the attraction to the individual who is not directly involved as victim, perpetrator or witness to the memorialised violence of the past can be explained as an inextricable mixture of three main universal human emotions. People are attracted to these sites through a mix of curiosity, empathy and, however distasteful, attraction to horror itself.

The unusual catches and holds people's attention. The spectacular natural phenomenon, built structure or event satisfies a human curiosity because it is out of the ordinary, daily familiarity. In so far as violence is unusual, it will therefore attract. Although we now regard gladiatorial combats, viewing mental patients (as in eighteenth-century London's Bedlam hospital), watching public executions or even treating battles as a spectator attraction (like Pierre at Borodino in Tolstoy's

War and Peace) as distasteful, the main motive was and remains the experience of the unusual. This motivation has much in common with 'disaster tourism', the viewing of the results of accidents or natural disasters. This may be seen as a distasteful intrusion on the sufferings of others which borders on the psychotically disturbed, or as only an expression of a seemingly universal insatiable appetite for witnessing the unusual.

Empathy relies upon the capacity of heritage consumers to identify themselves with the individuals involved in the memorialised violence. It is ethically more acceptable to experience an empathetic identification than to exercise mere voyeurism, although the distinction in emotion is difficult to draw and difficult to express through interpretation. Empathy is usually assumed to be with the memorialised victims ('this could have been me'). It could equally, however, be with the perpetrators of the violence ('I could have done that'), which raises highly sensitive issues of heritage interpretation. In particular, the creators and managers of the heritage of violence may justify graphic description as creating empathy with victims and thus preventing a recurrence of similar events. They have, however, little knowledge or control of the motives or reactions of consumers, who may be empathising with the perpetrators and may even be stimulated to replicate the violence. The idea of the visitor engaging in fantasy poses the questions: which fantasy has been selected, and which role is being acted out – that of the victim, or the perpetrator, or both?[9]

It may seem repugnant and morally unacceptable to be entertained by the accounts of the suffering of others. However, there has always been a link between descriptions of violence and entertainment. The portrayal of horror, once safely distanced or contained, and the evocation of emotions of fear and fascination has long been a staple of literature, folk stories and, more recently, film and television. The heritage of violence may be as entertaining as any of these media, convey similar emotions and be as morally acceptable. There may be a qualitative difference between an interpreted battlefield site or museum display and a tourist 'murder trail', 'ghost walk', 'chamber of horrors' and the like, but they all depend on the same exciting frisson of proximity to horror. The visit to the castle dungeon, the prison 'death row' and sites associated with genocide have a common element of attraction to horror. This, bizarre as it may seem, can be exploited as a major entertainment resource, which in turn raises the question whether this is the aberrant behaviour of psychologically unbalanced and socially anomalous individuals, or quite normal, as we are all 'children of the dark'.[10]

The motives of the producers

There are at least four main types of motive among those responsible for creating, preserving, maintaining, interpreting and managing heritage resources. These can be summarised as historical preservation, the evocation of empathy, fostering group identity and the shaping of better futures. Many of the professionals involved in the relating of the historical record and the preservation of the archives, artefacts and sites are uneasy about being ascribed external motives and may reject outright any suggestion that their work has any significance other than the objective relating of the factual historical record and the preservation, for its own sake, of associated artefacts and documentation. However, much interpretation of atrocity sites aims at stimulating empathy between the visitor and the memorialised victims. Visitors are encouraged to put themselves in the place of the innocent victim and feel similar emotions, thus bridging the gap between the past and the present. Success depends upon shaping a rapport between the victim and the visitor, often through the portrayal of mundane items and behaviour that evoke familiarity in the visitor. The personification of an otherwise anonymous group of victims could be termed the Anne Frank phenomenon, whose global success has been chronicled by Hartmann.[11] The personification of the Holocaust in a single girl both in the 'Anne Frank House' in Amsterdam and at Bergen-Belsen, where she and her sister died, has proved very effective in creating empathetic links between the visitor and the victimised group, especially among teenagers and young adults. Many atrocity site interpretations, including some of those from the Jewish Holocaust, attempt to move from the particular time-specific historical event to contemporary issues more directly familiar and relevant to the visitor: a widely encountered theme, well illustrated by the Jewish Museum in Berlin, moves from a persecution of Jews in history to current racial or ethnic intolerance. This both aids the visitor's engagement and, hopefully, identification with the victims and furthers a social policy agenda of using the past explicitly to influence the present and future for humanitarian goals.

An important use of all heritage, and the principal reason for its intentional creation by public authorities, is the fostering of group identity. Group cohesion, the identification of groups with places and the legitimation of ideologies and governments are dependent upon the acceptance and successful communication of a collective heritage. As heritage is one of the instruments by which social groups are formed, claims are laid upon places and authorities justify their right to rule, then atrocity heritage is a particularly effective means of achieving these objectives because of its powerful emotional appeal and its enduring memorability. Victimisation, in this sense of the consciousness of being a part of an actual or potential group of victims, both defines and strengthens group cohesion and also separates it from the outside 'other', who is at best a non-victim and therefore not us, and at worst a perpetrator of unfriendly actions against us. There is hardly a nation-state or political ideology that does not promote a 'freedom struggle' with martyrs and a demonised enemy. In the camps the question is often which victims are selected and who is the 'other'.

A very common ostensibly altruistic and philanthropic motive of those projecting the interpretation is the desire to avoid repetition of the atrocity. The didactic 'lest we forget' idea assumes that knowledge of past atrocities teaches valuable future lessons, especially the avoidance of repetition. Such attempts at pedagogic moralising have a poor historical record of success. It is almost impossible to know what lessons, if any, the visitor will draw from the interpretation. The visitor may draw lessons other than those intended, for example revenge rather than reconciliation. Remembering may exacerbate social and political division rather than fostering cohesion, which may be better served by collective amnesia. It could even be argued that such heritage interpretations may inspire rather than deter future repetition. The presentation of atrocity may anaesthetise rather than sensitise the visitor, making horror and suffering appear more normal and acceptable, rather than shocking and unacceptable. The relating of atrocity may destabilise susceptible individuals and the publication of especially horrific events show what could be done and thus inspire 'copy-catting'. With all of these motives, the significant point is that the motives and therefore messages of the heritage producers may be not only multiple, but also not the same as those of the consumers.

There is an often overlooked third group of participants, local residents, whose relationship to such heritage may well be ambiguous. These accidental participants in the sites and events commemorated may be gratified by outside interest, may profit economically from tourists or may resent the intrusion of outsiders, claiming their place. Current residents may not associate with the interpreted atrocity, being simply uninterested in it, or may even actively disassociate themselves from it as belonging to a different time or people. A simple explanation of lack of interest may be that the emotional experience of the visitor is temporary and too overwhelmingly powerful to be experienced continuously by residents. If current residents are associated with the perpetrators then alienation and resentment is likely. Those who live in Dachau, a pleasant residential suburb of Munich, may resent their worldwide strong place recognition. Óswiecim is a working, living Polish community as well as the location of Auschwitz.

Buildings and sites

It is important to remember that what are loosely termed 'concentration' camps include different functional variants. The timing and context of construction, the functional components and the ultimate fate of the buildings and sites differed accordingly: the death camps, now mainly in Poland, were created later and framed around the well-known means of extermination and were substantially obliterated in an attempt to destroy the evidence of genocide as the Soviet army advanced. The largest of these were the camps at Auschwitz-Birkenau. In contrast, the older concentration camps in Germany proper were not evacuated and destroyed by the Germans, no doubt in part because of the speed of the military collapse on the western front; and Bergen-Belsen, Buchenwald and Dachau (if less so Sachsenhausen) became iconic symbols of atrocity to the liberating and deeply shocked British and American military and, through cinema newsreels, their civilian populations.

In the surviving sites, many of the buildings were destroyed in the aftermath of liberation (the British army immediately burnt down Bergen-Belsen, ostensibly as a precaution against the spread of typhus) or in the process of subsequent reuse by successive regimes (both Auschwitz and Sachsenhausen were briefly used as accommodation by the Soviet authorities and then largely demolished in 1947 and 1950 respectively, while Buchenwald and others were used as denazification centres). Neither the Allied authorities nor the local populations had any immediate interest in preserving the sites. The main exception to this widespread demolition was Theresienstadt (Terezin) in Bohemia, which, being located in an existing historic fortress, has remained substantially intact.

Re-assembly or reconstruction of the camps is as inevitably controversial as with any other heritage site, confronting the opposing dogmas of the authenticity of the object with the authenticity of the visitor experience. Some reconstruction and some attempt to reassemble lost buildings has occurred in almost all the sites, but against a resistance to the creation of a tourism theme park. One small case that generated local media interest was triggered by a fire in a farm building in the northern Netherlands in 2009. The suspicion that the building had previously been located in the nearby Westerbork transit camp and had been acquired by the farmer after the war raised popular excitement, and it was asserted that this was 'barrack 57', which functioned as a workplace during the stay of Anne and Margot Frank in 1944. The tourism possibilities of a relocated and restored building with such associations evoked considerable enthusiasm, but expert doubts about authenticity and an unease about reconstruction led to a compromise of a 'symbolic reconstruction', whereby part of the building was placed upon its original site to give an indication of how it had been.

Visualising the conditions that once prevailed in the now largely vacant sites of Buchenwald and Birkenau, the chief location of the Auschwitz gas chambers, requires imagination from the visitor and a reliance on limited surviving structures and artefacts, partial reconstructions, subsequent on-site memorials and museum exhibits. Some convey powerful messages, notably the victims' belongings at Auschwitz, the crematoria at Buchenwald and the entrance gateways with their cynical slogans at both. Inescapably, perhaps, sanitisation compromises interpretation through the very clearance of human remains and so many structures. In Auschwitz the management of the natural environment, especially the fifty-eight remaining Lombardy poplar trees, provide a green backdrop to '*Arbeit macht frei*', which may appear to belie its dreadfulness. However, as some were planted by the authorities as concealment and some by prisoners, they have become part of the memorialisation (the so-called 'silent witnesses'), nevertheless raising a concern that the site becomes a manicured green park deflecting from its horrific historical reality.

Heritage interpretation over time

Tourism is a latecomer in a sequential evolution of the uses of such heritage, the most prominent being political legitimation of the regime in power in the various successor states, a marked divide occurring over time in at least three of the heritage interpretations involved. Since the principal victims of the death camps were Jews, the heritage of the camps has predictably been appropriated by the state of Israel, indeed being powerfully implicated in its very foundation. The Zionist return movement to Palestine was originally unsympathetic to European Jews and their supposed subservient ghetto mentality, unsuited to the assertive, optimistic utopianism of Zionism. However, the implementation of the 'final solution' and the plight of Jewish survivors of Nazism in 1945, who could not or would not return to their pre-1933 lives and locations, added urgency to their reception in Palestine and the creation of the state of Israel in 1948 as a safe refuge for Jews. Subsequently the term 'Holocaust' was appropriated not only to define the Jewish genocide but, ironically, to provide a retrospective validation for the political identity of the Israeli state, despite its original Zionist conception as the means to forestall such a catastrophe.

In Poland the changed political perspective occurred much later, with the end of the Cold War. The preceding Communist order had cast the genocide not as a uniquely Jewish but as an internationally victimising atrocity, in which Polish citizens as a proletariat had been the victims of a virulent derivative of capitalism. Not all of the victims of the camps were Jewish; some were victimised for their ideology rather than ethnicity; Jews were not the only ethnic group singled out for genocide; and many of the Poles murdered were Catholic. However, the presence of non-Jewish symbols, whether Christian or socialist, seemed to those who associated with the Jewish victims to deny their paramount role as victim. At Auschwitz the planting of a large cross (1979) and the establishment of a Carmelite convent (1984) led to protests from many visitors, supported by the state of Israel. Jewish demands for removal were met by local resistance in the form of the planting of many more unofficial crosses. The convent was moved in 1993 and the crosses were removed by the Polish authorities in 1999.

However, sensitivity about this heritage interpretation between visitors – many of whom associate with the Jewish victims, supported by sympathetic foreign governments – and the local Polish population, who may feel alienated from the heritage and even charged with co-complicity in the atrocity, remains.[12] Charges of anti-Semitism and 'anti-Polonism' were exchanged. Such sensitivity re-emerged in 2007 when UNESCO, at Polish insistence, changed the name of the world heritage site designation from 'the Polish death camps' to 'German Nazi concentration and extermination camps'. The immediate reactions in Poland and around the world to the theft of the famous sign in 2009 and fears for the consequences of a revival of animosities were only forestalled by its fortunate rapid recovery and the seemingly non-ideological motivation for the crime.

In Germany a parallel but more nuanced post-Cold War re-conceptualisation of the atrocity heritage has occurred, with respect primarily to the former DDR. The Communist-led state had cast the Nazis as a group rather than Germans as a people as the perpetrators and identified them with those forces still dominating the West German state. It had cast the communists as leaders of the international resistance to National Socialism and had centred this narrative upon Buchenwald, where Communist leader Ernst Thälmann had been imprisoned and murdered in 1944. This interpretation had extended to Buchenwald's self-liberation by prisoner resistance before the arrival of US forces in 1945. The unification of DDR and BRD in 1990 created a delicate problem of re-conceptualisation, which could neither dismiss the DDR perspective nor entirely deny its charges against the West. This was gradually resolved through the 1990s in Buchenwald (and similarly in Sachsenhausen and elsewhere) after improvised management and interpretation changes by decentralising this re-conceptualisation to autonomous but better research-financed place-specific agencies. These in Buchenwald recast the museum interpretation of its development as an evolutionary process from the camp's origins and Nazi history, through its role as a denazification camp from 1945 to 1949 (in which many also died), through the DDR's resource-poor, politically motivated interpretation to a more comprehensively researched understanding consistent with a liberal, pluralist democracy within an integrating Europe in a new century.[13] Continuities remain, however; notably, memorials to international victim groups and the DDR's Buchenwald warning and memorial tower, which still dominates the hillside symbolically overlooking Weimar.

The DDR's interpretation and marketing of Buchenwald had expressly linked it to Weimar, reflecting the Nazi glorification of this focus of classical German culture. This prompted the construction of a nearby concentration camp as a source of slave labour to maintain it. The original name, 'Ettersberg', was changed to Buchenwald ('Beechwood') for its romantic rusticity, and it contained the remains of the renowned 'Goethe's oak'. The DDR co-marketed Buchenwald and Weimar as encapsulating what it saw as the German contradiction, which has been continued by democratic united Germany, although differently nuanced. The communists are no longer exclusively lionised as both resistance leaders at Buchenwald and the true inheritors of what is best in German culture, as embodied in Weimar's classical pantheon, personified by Goethe and Schiller and more recently and ambiguously the Bauhaus school of design. Moreover, democratic Germany accentuates the nobler heritage of Weimar in its nostalgia for the lost Weimar Republic founded there and extinguished by Nazi–Communist polarisation in the 1930s. Weimar's European City of Culture initiative in 1999 highlighted the historical German contradiction with the designation of a time transect (*Zeitschneise*) path linking a classical palace in Weimar to the Buchenwald site. The ambiguities over approaches to the Nazi past remained, however, and are illustrated by the Nazi *Gauforum* administrative complex, which is preserved, renovated and reused but only very discreetly marked and largely un-interpreted.

The juxtaposition and, indeed, original deliberate association of Buchenwald and Weimar prompts an exploration of the idea of co-marketing Holocaust-related tourism experiences and promoting networks of sites. Such packages may consist of different Holocaust elements, such as the 'Auschwitzland' product that combines the 'Schindler experience' in the renovated former Jewish ghetto of Krakow-Kazimierz, and more recently the Schindler factory–museum in Krakow-Podgorze.[14] Equally, Holocaust and non-Holocaust elements may be combined where locations permit in a varied tourism product. Buchenwald/romantic Weimar and Auschwitz/baroque Krakow are obvious examples of seemingly incongruous elements in which the camps become day excursion possibilities within a broader holiday experience.

As noted above, there is another, often overlooked, dimension to at least some of the camps, namely their occupiers after the defeat of Nazi Germany. Although many camps were destroyed after liberation, some were not and took on new functions. Theresienstadt was used to house expelled Sudeten Germans and Dachau was a refugee centre from 1948 to 1965. The transit camp of Westerbork in the Netherlands was neither constructed by the German authorities nor disused on liberation. It was built by the Dutch government, ironically to house pre-war refugees from Germany. In 1945 it was used to intern collaborators and from 1950 to accommodate the demobilised Dutch East Indies colonial army (KNIL), largely from the South Moluccas and their families, until 1970 when the camp was finally dismantled. The heritage issue is: how should such sequential occupancies be interpreted? Which groups should be memorialised and which marginalised?

Concluding questions

This chapter concludes with recognition that more questions have been raised than answers proffered, but only reflects the inherent complexity involved in relating tourism to the camps. The Holocaust tourist is not an easily defined and homogenous consumer of a standardised tourism experience, with predictable motivations and behaviour patterns. Tourists may have travelled far, with the camp as the ultimate destination of a personal pilgrimage, or equally may be holiday excursionists indulging a spontaneous general heritage interest. Visits may be motivated by profound personal conviction, by the collecting of pre-marked, 'must-see' attractions, by a vague curiosity about a distant historic event, or may even be involuntary, as in school educational visits. The tourist may identify and empathise with the victims, experiencing emotions of grief, sorrow or anger; feeling guilt, shame or remorse, or just curiosity at remote, bizarre and inexplicable human behaviour. The practical consequence of such diversity of expectations is that the packaging, promotion and on-site consumption behaviour raises management issues of peculiar complexity and sensitivity.

There are also questions relating to the producers, the product being sold and the management of the site. The objectives of the producers are as varied, and often as contradictory, as the consumers. Site managers responsible for the presentation of the tourism experience typically see themselves as motivated by philanthropy, altruism and humanitarianism, and feel obligated by a strong pedagogic moralising mission. Although tourism is in essence an economic activity pursued for commercial reasons, few site managers would list economic gain or the financial contribution to local economies as one of their main objectives. Although most site managers would accept that tourism is an important justification and validation of their endeavour and provides the audience for the lessons being taught, few would claim that tourism is the reason for the existence of the heritage site and main motive for their labours. The camps, like all heritage attractions, are in multiple use and the reception of visitors is only one of these; the camps would continue to exist as heritage sites, with research, education and just existence values, without tourists. This raises both general and specific questions for the producers. How are the various objectives of the heritage producers conveyed in the messages transmitted, and how are these messages received by consumers? Are the messages that are being projected those that are received, and is the tourist consuming the same product experience that the producers think they are producing? In terms of on-site management, how should tourist behaviour be regulated and influenced so that the diverse experiences can coexist?

There are also questions relating to the wider impacts of atrocity heritage tourism upon the societies of both hosts and guests. The interpretation of the Holocaust serves a number of wider political concerns, including political and ideological legitimation of governments and their dominant values; the socialisation and social cohesion of the group, even to the extent of nation-building; the satisfaction of a desire for the contemporary stewardship of historical sites, artefacts and memories as a legacy for bequest to an unspecified future. The camps are not unique as atrocity heritage sites and atrocity tourism attractions. The Jewish Holocaust is not the only genocide to have happened, to be currently memorialised and sold on tourism markets. The camps do serve, however, to sharply focus the more general issues raised above. The scale, completeness, relative recentness and visibility of the camps, together with more pragmatic factors such as their location in a region of high accessibility and high tourism demand, all ensure that they will be visited in ever increasing numbers, and that their interpretation will continue to evolve.

Notes

1 J. E. Tunbridge and G. J. Ashworth, *Dissonant Heritage: The Management of the Past as a Resource in Conflict* (Chichester: Wiley, 1996).
2 G. J. Ashworth, "The Memorialisation of Violence and Tragedy: Human Trauma as Heritage," in B. Graham and P. Howard, ed., *The Ashgate Companion to Heritage and Identity* (Aldershot: Ashgate, 2008), 231–44.
3 United Nations, *Convention on the prevention and punishment of the crime of genocide* (1948).
4 G. J. Ashworth and R. Hartmann, *Horror and Human Tragedy Revisited: The Management of Sites of Atrocities for Tourism* (New York: Cognizant, 2005).
5 M. Foley and J. J. Lennon, "JFK and Dark Tourism: A Fascination with Assassination," *International Journal of Heritage Studies* 2, no. 4 (1996): 198–211; J. J. Lennon and M. Foley, ed., *Dark Tourism: The Attraction of Death and Disaster* (New York: Continuum, 2000).
6 A. V. Seaton, "Guided by the Dark: From Thanatopsis to Thanatourism," *International Journal of Heritage Studies* 2, no. 4 (1996): 234–44.
7 R. Hartmann, "Dark Tourism, Thanatourism, and Dissonance in Heritage Tourism Management: New Directions in Contemporary Tourism Research," *Journal of Heritage Tourism* 9, no. 2 (2014): 166–82.
8 G. J. Ashworth and R. K. Isaac, "Have We Illuminated the Dark? Shifting Perspectives on "Dark" Tourism," in *Tourism Recreation Research* 40, no. 3 (2015): 316–25.
9 G. Dann, "Tourism Motivation: An Appraisal," *Annals of Tourism Research* 8 (1981): 187–219.
10 G. Dann, "Children of the Dark," in *Horror and Human Tragedy Revisited: The Management of Sites of Atrocities for Tourism*, ed. G. Ashworth and R. Hartmann (New York: Cognizant, 2005), 233–52.
11 R. Hartmann, "The Anne Frank House in Amsterdam: A Museum and Literary Landscape Goes Virtual Reality," *Journalism and Mass Communication* 3, no. 10 (2013): 625–44.
12 A. Charlesworth, "Contesting Places of Memory: The Case of Auschwitz," *Environment and Planning D: Society and Space* 12, no. 59 (1994): 579–93; W. Miles, "Auschwitz:

Museum Interpretation and Darker Tourism," *Annals of Tourism Research* 29, no. 4 (2002): 1175–8.
13 J. Beech, "The Enigma of Holocaust Sites as Tourist Attractions – the Case of Buchenwald," *Managing Leisure* 5, no. 1 (2000): 29–41.
14 G. J. Ashworth, "Holocaust Tourism and Jewish Culture: The Lessons of Krakow-Kazimierz," in *Tourism and Cultural Change*, ed. M. Robinson, M. Evans and N. Callaghan (Newcastle: Channel View, 1996), 363–7.

6 Guilty landscapes and the selective reconstruction of the past

Dedham Vale and the murder in the Red Barn

Martin Spaul and Chris Wilbert

Prologue: the decline of an English murder

In July 1828 the journalist James Curtis travelled from London to the Stour Valley in Suffolk. His impressions on arrival centred on the rain of a wet summer, as the flooded Stour carried away the recent hay harvest and with it the 'hopes of the husbandman'. The sight drew expressions of concern from Curtis, as he reflected that 'many humble cottagers' had been deprived of that which might have carried their solitary cow through the winter.[1] Further reflection on his part might also have elicited fear, as the loss of a significant part of the harvest was not usually borne stoically by poor agricultural labourers, who perpetually lived on the borders of starvation, anger and violence. East Anglia in the early nineteenth century was marked by a continual climate of violence.[2] In 1816 – the 'year without a summer' – the failure of the harvest led to food riots, machine-breaking and incendiarism as the rural labourer 'found himself caught between unemployment and scarcity'.[3] A similar pattern of events was repeated throughout the subsequent decades.

Curtis' main concern on this journey was, however, not with collective, public violence but with its individual, clandestine forms; and his profession was that of crime reporter. Making his way to the village of Polstead, he duly noted the picturesque rurality of the village, set on a hillside and affording 'a beautifully romantic prospect' with its gleaming white cottages, church spire, spacious fields and an imposing country seat reassuringly 'built in Portland stone'. The object of his visit became apparent as he was moved to invoke the terms of pastoral tragedy: 'is it possible that, in the midst of this little Eden, a village swain has imbrued his hands in the blood of his damsel?'[4] Although this characterisation was congruent with the prevailing mythologies of harmonious rural existence, they were oddly inadequate to the realities of rural life and the brutality of the events he had come to report.[5] His visit was occasioned by the discovery, in April 1828, of the body of Maria Marten, the daughter of the village mole-catcher, in a shallow grave dug into the floor of Polstead's 'Red Barn', a local landmark. Marten had been killed the previous May by William Corder, her erstwhile lover and a local gentleman farmer, but the crime only came to light with the chance exhumation of the body. Curtis, experienced in satisfying the public thirst for sensational crime, had come to flesh out his story with 'human interest' before travelling to Bury St Edmunds to report on Corder's trial and, as it transpired, execution.[6]

By the time Curtis had arrived in Polstead, a large-scale dark tourism industry had been in operation for several months, as the case had been an immediate nationwide sensation. His account traced the sites that had already become the principal stations in the tourists' emotional recreation of the tragedy. The focus of attention was the Red Barn and its shallow grave, although perhaps journalistic hyperbole was at work in Curtis' estimate that 'since the time the murder was first discovered . . . on a moderate calculation, two hundred thousand persons have been under its roof, many of whom have travelled fifty miles and upwards'.[7] Further sites on the tourist pilgrimage offered the opportunity for Curtis' moralising reflections: Maria Marten's humble family cottage, standing testimony to the rural myth of the poor-but-honest village maiden; Corder's imposing house, the unaccountable origin of spontaneous evil; and the tragic finality of Maria's last resting-place in the village churchyard. The flood of tourists continued long after the trial and into the following decades, as the red-blooded dark tourism practices of the nineteenth century stripped the boards of the barn for the manufacture of souvenirs ranging from toothpicks to snuff boxes and consumed Maria's gravestone in fragmented keepsakes.[8] The remnants of the Red Barn burned down in 1842; as Jones documents, the 1840s were a period of intense incendiarism in East Anglia.[9]

Some two hundred years later, a tourist following Curtis' route into the Stour Valley is confronted with a landscape in which many of the outward features of the 1820s remain intact, although these have been drawn into thoroughly modern systems of meaning. The village of Polstead is a designated conservation area with more than twenty listed buildings, including Corder's house and Maria Marten's cottage; but the reasons for this heritage status are exclusively architectural and topographical. The conservation area appraisal for Polstead makes no mention of its dark past, and Maria Marten's cottage is listed for its qualities as a timber-framed building of the seventeenth century.[10] Polstead is also drawn into a landscape of wider significance, as it lies within parts of the Stour Valley drawn into the Dedham Vale Area of Outstanding Natural Beauty (AONB), popularly marketed as the 'Constable Country' tourist destination and valued for its preservation of the views enshrined in the paintings of John Constable. The features that Curtis listed in his evocation of a 'little Eden' make up much of the vocabulary of Constable's paintings, contributing to their quintessential Englishness.[11] The framing of Polstead as part of a wider heritage landscape experience has not entirely occluded its status as a dark tourism site: it is promoted by the Automobile Association, with the Red Barn murder providing a historical curiosity to enliven a circular 'pub walk', and the Polstead community shop sells a locally produced guide which enables the visitor to take a village tour to the accompaniment of the details of the murder. As Lucy Worsley noted, the local community would prefer not to remember its dark past; this impression was reinforced by the mild exasperation met by the present authors when purchasing a copy of the local guide.[12]

Guilty landscapes and the erasure of the past

Over a period of two centuries, it may be seen as unsurprising that the role of the Red Barn murder in defining the 'place image' of Polstead and the Stour Valley has declined to that of a background curiosity.[13] However, this process of forgetting has involved more than the loss of a single event. The individual act of violence that ensured Polstead's notoriety throughout much of the nineteenth century took place against a background of wider class conflict and collective violence that characterised rural life in East Anglia and, just as the landscape of Stour Valley has forgotten the Red Barn murder, it has also forgotten the starvation, bread riots, rick burnings, hangings, transportation and imprisonment that comprised the common experience of the rural poor. The murder seems to have been perceived by contemporaries as an index of these wider events, and more than an appetite for sensationalism seems to have been at work in the immediate notoriety of the Red Barn murder. An important factor in the rhetorical power of the Red Barn murder was that captured by Curtis: the shattering of the popular and assiduously promoted ideology of the rural, that rural society was more harmonious, happy and virtuous than urban society.[14] More specifically, the 1840 melodrama *Maria Marten, or the Murder in the Red Barn*, produced in London at the Marylebone Theatre, was immensely popular with the theatre's working-class audiences. This popularity is partially accounted for by a plot that reversed the stereotype of the threatening, disorderly poor, depicting a 'poor wronged girl' as the victim of a self-serving middle class – a significant symbol in a period marked by the punitive 1834 Poor Law.[15] The Red Barn murder takes on a multi-layered significance when seen against the backdrop of a society 'shot through, from top to bottom, with the dread of some wild outbreak by the masses that would overthrow the established order'.[16]

The erosion of the memory of these events cannot be seen as inevitable. Not all landscapes forget their dark past, even at a distance of centuries; for example, the Highland Clearances still play a vivid – intensified and mythologised – role in defining the place image of the highlands of Scotland.[17] It is also significant that, while the Red Barn murder may play little active part in the place image of the Stour Valley, it has an active after-life in broader traditions of popular culture disconnected from place.

To set up a theoretical framework for analysing the acts of forgetting performed by the landscape of the Stour Valley, it is useful to begin with the figure of the 'guilty landscape' introduced into tourism studies by Reijnders, who used the term to characterise beautiful landscapes into which fictional murders had been introduced, and which had hence become tourist attractions. Such landscapes trade on the shock value of the incongruity between beauty and cruelty (a trope understood by James Curtis), with crimes embedded in a landscape where 'natural beauty was so luxuriant that it seemed impossible that murder and torture could have taken place here'.[18] The original formulation of the concept of the guilty landscape, by the Dutch artist and writer Armando, is more tentative and ambiguous. He used it to capture the kind of guilt that attaches to a landscape – now silent, flourishing and beautiful – that has witnessed terrible events in the past. It provided a means of articulating his feelings about the forested landscape around Kamp Amersfoort, a Nazi concentration camp that is now a Dutch national monument.[19] Writers on the Holocaust have connected Armando's 'guilty landscape' with the broader problems of testifying to a receding past and the necessity of keeping the memory of atrocity alive, with van Alphen locating the guilt of the landscape in its 'refusal to testify'. The dissembling beauty of a guilty landscape hides or camouflages the crimes of the past, and the guilt of the landscape is that of the witness who has chosen to remain silent.[20]

Chrisje Brants has noted that the concept of a guilty landscape – as formulated above – is problematic as the basis for a more systematic analysis of the responsibility for forgetting a dark past. The absence of any overt recognition of human action, with agency directly attributed only to the landscape, means that a guilty landscape serves only as a metaphor for a broader, incompletely specified, human guilt.[21] For Reijnders' dark tourism sites, the 'guilty landscape' label serves to designate a particular kind of tourist experience, but it does little to explain how the possibility of that experience has arisen. Similarly, the observation that the Stour Valley has forgotten its dark past offers no explanation of why this landscape happens to be experienced as harmonious, restful and beautiful by its visitors, why selected aspects of its history serve to define its dominant meaning in the present day and why a past clouded with cruelty, violence and suffering plays little or no part in this meaning.

Answering these questions requires a more articulated conception of what constitutes a 'landscape', and also of how a landscape can act as a vehicle for memory. Although the concept of landscape is a complex construct,[22] the Stour Valley's status as a 'cultural landscape' under the European Landscape Convention provides a convenient shorthand definition suitable for the current analysis.[23] This analysis of landscape may be combined with models of cultural memory derived from the work of Assmann to derive an outline account of how a landscape may refuse to testify to its past.[24]

A cultural landscape may be analysed into two principal components, theorised in two different traditions of study.[25] On the one hand, a cultural landscape is formed by the interaction of a human community with the forces and resources of nature; from this perspective, a landscape is an area worked for human purposes. On the other hand, a landscape is also an object of perception, of appreciation and evaluation by people; a perception conditioned by the cultural preferences and expectations brought by those who gaze on a 'landscape of views'. The cultural landscape of the Stour Valley is simultaneously the product of an agrarian past – a working landscape – and the cultural cachet of an idealised version of the landscape captured by Constable at a single point in that past. The current form of the landscape is the outcome of interactions between the scenic value placed on it by visitors and the hybrid agricultural and tourist economy that sustains its physical characteristics. What aspects of the past are held by the landscape – in its physical forms, its tourist interpretation and the perceptions of its visitors – are the outcome of a selective evolution conditioned by both the working landscape and its popular cultural reception.

This process of selective landscape evolution may be linked with established models of cultural memory. 'Cultural memory' is used by Assmann to designate the forms of long-term memory that operate in a society after living, word-of-mouth, 'communicative' memory has died out.[26] The long-term memory of a society is held in diverse mediated forms – texts, monuments, architecture, landscapes, and so on – and rehearsed in 'sites of memory', of which heritage landscapes and their associated centres of interpretation are specific instances. The holding and rehearsal of memory is a dynamic process in which cultural materials from the past are re-worked according to the perceptions and priorities of the present – a process which enables selective forgetting.[27] Connerton argues that forgetting is central to cultural memory, and his typology of forms of forgetting provides a framework for explaining the means by which a guilty landscape may refuse to testify.[28] 'Benign repressive erasure' – marginalising cultural memories that cannot be contained within a dominant narrative – is typical of institutionalised heritage sites that attempt to present their visitors with orderly versions of the past. The Stour Valley – managed and interpreted to reflect its association with the work of Constable, and the harmonious, productive landscape he depicted – cannot easily accommodate within its untroubled narrative the disorderly facts of poverty, violence and injustice that marked the working lives of those who produced the landscape. Part of the difficulty of accommodating alternative cultural memories may be attributed to another of Connerton's categories, that of 'structural amnesia'. It is only possible, within a site of memory, to hold aspects of the past that are capable of being represented with the expressive means available at the site. The Stour Valley is predominantly a recreational landscape in which the gentle leisure pursuits of walking, cycling and sightseeing are carried out against a comfortably scenic backdrop. The cultural memories that can be held in such a site are those that can be attached to historic views and buildings, or held within guides compatible with the dominant recreational uses. Nuanced historical narratives cannot easily be expressed.

With the help of Ann Rigney's distinction between sites of memory that are locatable places – museums, archives, heritage centres, historic landscapes, and so on – and those that are virtual spaces defined by the circulation of media objects, it is now possible to formulate a general thesis concerning 'guilty landscape' dark tourism experiences.[29] A landscape that is guilty in van Alphen's sense – refusing to testify to its dark past – is a site of cultural memory in which repressive erasure and structural amnesia have ensured that the dark past is no longer represented at the site. It can be perceived as guilty, and thus provide the shock experience identified by Reijnders, only if there are other sites of cultural memory in which the dark past of the landscape has been preserved. In the case of the Stour Valley, the guilt of the landscape lies in its forgetting of the individual and collective violence of the early nineteenth century, and the perception of that guilt is possible because virtual sites of memory have preserved the past that the landscape denies.

The Stour Valley as guilty landscape: landscape and forgetting

The role of John Constable and his paintings of Dedham Vale in defining a specifically English version of rurality has been extensively studied, as has the development of 'Constable Country' as a tourist destination.³⁰ The development of Constable's approach to landscape art, and its relationship to his native region, has received similarly intensive treatment.³¹ For the purposes of this study, it will only be possible to trace a single line of development through this established literature, relating it to the way in which the Stour Valley has been framed as a heritage landscape, and the conservation and interpretation regimes currently in place. This line of development will show a continuity in a 'form of forgetting' that has been in operation for nearly two hundred years, suppressing the violent actuality of the lived experience of the common people of the Stour Valley – those who shaped its now valued qualities – and contributing to the near-invisibility, in the official forms of interpretation for the area, of the darker episodes of its past.

The Dedham Vale landscape is viewed as culturally significant because of its association with Constable and the features that he highlighted: the 2010–15 management plan lists the features of the river, the farmed landscape and the historic settlements and emphasises the visual continuity between Constable's images and the views available in the present day.³² The landscape was designated as an AONB in 1970 to protect this visual continuity from pressures for housing development, so that the area would still be available for 'quiet enjoyment' and its historic environment conserved. Today's institutionalised conservation is thus rooted in the visual selection and emphasis that marks his iconic paintings. The artistic and literary heritage of the romantic era was subject to extensive academic scrutiny in the 1980s and 1990s, with the emergence of forms of criticism based on contextualising art works within the social and political conditions that surrounded their production.³³ The application of this criticism to the central figures of English romanticism (in particular Constable and Wordsworth) has produced insights of significance for the ways in which English cultural landscapes are viewed and used in the present day.

Studies by John Barrell and Ann Bermingham applied this critical procedure to the work of Constable, using different contextual premises to reach broadly similar conclusions.³⁴ Barrell noted that Constable presents us with a serenely civilised landscape, but for the most part pressed those who created the landscape – both the labouring poor and their privileged masters – either into the background of his images or somewhere outside of the frame. If people do intrude into one of Constable's paintings, then they are represented indistinctly 'to evade the question of their actuality'.³⁵ Constable's landscapes abound in detail that carries human associations – from the fine detail of 'old rotten planks, slimy posts, and brickwork' to 'villages, churches, farm houses and cottages' – but those human associations are not permitted to extend to fully individuated people.³⁶ When people do appear in his paintings, they are often reduced to blurs of colour, engaged in some task of shaping or working the landscape. As Bermingham observes, the figures are absorbed into the landscape in such a way as to suggest that they are oblivious to the spectacle of nature being rendered by the painter.³⁷ The blurred figures labour to produce the spectacle framed in the painting, while the viewer is placed in the 'outsider' subject position which is the defining characteristic of landscape for Williams.³⁸ Both Barrell and Bermingham concur that the 'human beings' (reduced as they are) in Constable's paintings act as blind automata in a system of production – a system of production represented by the hedged fields, the locks and quays, the mills and the barns that defined 'Constable Country' in the early nineteenth century and continue to do so in the present day.[39] Barrell and Bermingham offer complementary accounts of the roots of this selectivity: for Barrell, the distancing of the agricultural workforce may be traced to Constable's political conservatism, which grew virulent in the agrarian uprisings of the 1820s;[40] for Bermingham, Constable's visual language is the result of a psychological need, rooted in the details of his troubled biography, to push the actuality of East Bergholt's society (both high and low) from his canvases. Whatever the reason for his selective vision, it engendered a visual language broadly characteristic of romantic representational practices: a 'language of the heart' that displaced 'the language of society'.[41] In this visual language a worked landscape became a piece of nature, and painting – allied to the discipline of close observation – became an activity akin to natural history.

As Urry has argued, certain aspects of modern western tourism are derived from the romantic mode of an encounter with landscape dominated by vision.[42] A conception of landscape in which the human producers of the landscape are effaced into a distant and anonymous shaping force, on a par with the anonymous forces of nature, has had a powerful legacy; it has shaped both the legal frameworks through which landscapes are individuated and managed and the academic disciplines used to conceptualise landscapes. Selman and Swanwick show how romantic conceptions of landscape, championed by Morris and Ruskin, were influential on the preservationist movements of the nineteenth and early twentieth centuries, with a 'preoccupation with landscape as scenery and a somewhat escapist emphasis on aesthetics, picturesque views and a Romantic construction of nature . . . pre-eminent in influencing legislation'.[43] These institutionalised formations, and the ways in which they have been taken up in tourism promotion, media and marketing, have helped form both particular ways of looking or gazing on 'Constable Country' and the management of the cultural product itself.

The result of this visual preoccupation was that cultural landscapes – as they would now be termed – were characterised and valued for their 'natural beauty', a phrase that signalled a view of human labour (and those who perform it) as on a par with natural processes. Studies of the academic disciplinary assumptions of both geography and landscape archaeology have shown how the romantic tradition of the close observation of landscape features (in the manner of Constable and Wordsworth) has sanctioned forms of study that push into the background the social and political concerns of both past and present: of the past, by not exploring fully the socio-political processes at work when a landscape was shaped and hence what the landscape meant to those involved; of the present, by failing to explore the ways in which our current socio-political preoccupations determine our priorities for understanding and conserving landscapes.[44] These exclusions have not gone unchallenged in current critical approaches to geography or landscape archaeology.[45]

Against this background, it becomes possible to understand how some of the recent practices in the conservation and interpretation of the Stour Valley function as a process of selective remembering and forgetting. 'Managing a Masterpiece' was a project supported by the Heritage Lottery Fund; between 2011 and 2014 a series of studies and public engagement exercises were carried out to document and disseminate the heritage of the Stour Valley.[46] The terms in which these studies were carried out were consistent with both the general significance of Areas of Outstanding Natural Beauty – a focus on the natural environment, and the visual dimension of human settlement and land use – and the specific visual heritage marked out by Constable (and other artists) in the Stour Valley. The project activities are set out in Table 6.1 below.

The conception of what constitutes Stour's cultural landscape and the way in which such landscapes are to be understood and conserved are key determinants in framing the conservation projects of 'Managing a Masterpiece'. The dominant emphasis is on *what* has been produced by historical processes, with the question of *how* the landscape has been produced only conceptualised through patterns of work and use. The aspects of *how* the landscape has been produced that might lead to an enquiry into the history and experience of the individuals enmeshed in the process – the landscape as the outcome of a fully articulated way of life – do not obviously appear in the programme. This is not a failing of the project, or the result of some kind of oversight; the project is deeply embedded in the ways in which cultural landscapes are conceived in the defining legislation and the academic disciplines associated with historic landscape. The forgetting of the lives of those who made the landscape is institutional.

Table 6.1 The 'Managing a Masterpiece' Project: Project Outline (abstracted from Stour Valley Landscape Partnership, n.d.)[47]

Sub-Projects – Understanding the Masterpiece	
Landscape lessons: an outreach programme on the management of landscape, historical features and heritage.	Historic landscape study: a survey of landscape features and character.
Built history: a survey of built heritage.	Slimy posts and brickwork: a survey of the river and navigation structures.
Hidden history: archaeological survey focused on the historic landscape.	Stripping back the layers: community workshops in conservation.
Medieval masterpieces: conservation and interpretation of medieval monuments.	
Sub-Projects – Conserving the Masterpiece	
The grain of the canvas: management and conservation of natural heritage.	Painter's views: enhancement and restoration of views recorded in art works.
Handing on heritage skills: conservation training programme.	The 'John Constable': the restoration of a Stour river barge.
Sub-Projects – Celebrating the Masterpiece	
The view from the tower: recording the views from church towers.	Raising the profile of the valley: raising public awareness.
Increasing access not traffic: sustainable transport.	New masterpieces: art workshops and exhibitions.

Remembering individual and collective violence

The second half of the 'guilty landscape' equation is the preservation of the memory of a dark past in sites other than those that define a cultural landscape, and the heritage tourist encounter with it. For the Red Barn murder, the sites are located principally in the spaces of popular culture. For the social context of the murder, and the troubled nature of the agrarian society in which it took place, the sites are principally those of branches of academic history focused on the experience of common people. Both of these categories of site have their own characteristic forms of remembrance, and have limitations on the kinds of cultural memory that can be held and the audiences that engage with them.

The Red Barn murder made its way into the public consciousness – a popular public consciousness with an insatiable demand for the lurid details of violent crime – through the medium of broadsheets such as those published by the Catnach Press.[48] These broadsheets dressed less than factual accounts of the murder and Corder's execution in lurid terms for the prevailing public taste, with the volume of sales sending Catnach on his progress 'from a poor man to a gentleman'.[49] Quite apart from Catnach's ability to embellish and sensationalise the accounts of the crime, the Red Barn murder came ready-made to enter popular mythology, simultaneously instantiating a wide range of tropes active in tales of violence. The 'murdered sweetheart' had been a popular theme in ballads since the mid-seventeenth century, and the 'plot' of the Red Barn murder matched a common template: a naïve young woman lured by her seducer with promises of marriage and murdered in some lonely place.[50] As told by Catnach, the events of the Red Barn murder also involved the discovery of the body after a revelation in the recurring dreams of Maria Marten's stepmother; this immediately connected the murder with a long history of songs and tales in which evil deeds are brought to light by supernatural means.[51] The popular press' portrayal of Maria Marten as a local beauty also tapped into another powerful cultural motif (still active in crime fiction) of the beautiful female murder victim.[52] Driven by the rhetorical power of these elements, the retelling of the Red Barn murder span out into a wide variety of media, and the story was morphed and mythologised as it travelled. The 'Suffolk Tragedy' broadside ballad, which retained a semblance of factual detail, was recovered in 1972 as a 'derivative folk song' stripped of the specifics of time and place.[53] 'Corder's confession' – doggerel verse invented by Catnach for a broadsheet recounting Corder's execution – surfaced in an 'electric folk' version recorded by the Albion Band in 1971. Most notably, the murder became the subject of a melodrama tradition which is the basis for popular revivals to the present day (including a 1935 Hollywood film and a 2012 rock opera performed in Los Angeles).[54] The murder was even celebrated in the form of a Staffordshire pottery set of the barn and accompanying figures.[55]

The transmission of the Red Barn murder involves its own kind of loss and forgetting; as the tale spread through the paths of popular tradition, specificity and context were eroded. A real event, set in a troubled agrarian landscape of class conflict, took on a status somewhere between fact and fiction. It is in the nature of traditional transmission that elements of an event are taken from their context and made to bear meanings that typify the tradition.[56] This force parallels that which drove real people and events from the paintings of Constable, who was obliged – in order to protect the visual rhetoric that he had developed – to deny in his paintings a reality that he acknowledged in personal correspondence. The landscape that provides the models for the conservation of the Stour Valley were, in part, painted in an environment where there was, as Constable describes in a letter from 1821, 'never a night without fires near or at a distance', and serene views of Stoke-by-Nayland church – a landmark visible from Polstead – were also of a site in which the house of the local landowner was attacked by a violent mob in 1815.[57] The shared social context for both the Red Barn murder and Constable's work in the Stour Valley is held in a different site of memory: academic history's 'history from below'. This historical tradition focuses on the representation of the experiences and environments that define the 'ordinary lives' forgotten by mainstream political history. Since the formation of the 'history workshop' in the 1950s and the publication of E. P. Thompson's *The Making of the English Working Class* in 1963, specialist histories of working communities – including those of rural communities and their discontents – have provided a means of understanding aspects of common experience clouded by mythologies and misrepresentations.[58] These revisionist views of some of the 'sacred spaces' of the English imaginary have not always been welcome, and in 2011 Rosenthal detailed the resistance to attempts to integrate aspects of 'history from below' into the public understanding of the Stour Valley.[59]

Dark tourism and human lives: an ethics of remembering

Lucy Lippard has argued that 'The underlying contradiction of tourism is the need to see beneath the surface when only surface is available'.[60] For certain forms of dark tourism very little in the way of surface is available, and yet there are still poignant experiences to be had. A battle site may be no more than a field, a place of dispossession no more than a heap of stones, but what makes the experience is narrative recreation – a virtual, literary or historical site of memory reaching out to a place, and creating a new site of memory. What is problematic about the guilty landscape of the Stour Valley is its resistance to the intrusion of dark narratives that have perhaps a greater claim to a home in the landscape than Constable's scenes.

Oosterman, writing of 'guilty landscapes' and trying to take a more constructive way forward with the guilt of such landscapes, argues that 'Guilt has been effectively used to control and manipulate the masses. But it can also be the start of a change for the better: awareness, concern, action.'[61] The disruptive potential of the Red Barn murder is present in its longevity and popular cultural reach; its limitation is the way in which popular culture has preserved it, as a dramatic narrative cut adrift from the substance of the social order in which it originated. The fullest account available to the tourist, in Polstead's local guide, is meticulous in its 'local history' detail: information from parish registers, family trees, local land ownership, woven into a detailed reconstruction of the narrative. For a tourist to realise that the murder can bear a wider significance, and for it to provide an inkling that all was not well in Suffolk in the early nineteenth century, a connection would have to be made with other sites of memory largely only available to those with a determined research agenda. The present owes a duty of remembering to the nameless lives of the past – a duty that might be carried out in the guilty landscape of the Stour Valley by creating a well-informed and contextualised background to a dark tourism experience that has lost its original force.[62]

Notes

1. J. Curtis, *An Authentic and Faithful History of the Mysterious Murder of Maria Marten* (London: Thomas Kelly, 1828), xi.
2. Peacock, cited in P. Muskett, "The East Anglian Agrarian Riots of 1822," in *Agricultural History Review* 32, no. 1: 1–13.
3. M. Meisel, *Realizations: Narrative, Pictorial, and Theatrical Arts in Nineteenth-century England* (Princeton: Princeton University Press, 2014), 1.
4. Curtis, *An Authentic and Faithful History* xii, 50.
5. See C. Payne, *Toil and Plenty: Images of the Agricultural Landscape in England 1780–1890* (New Haven: Yale University Press, 1993), 24.
6. For recent, and more complete, accounts of the details of the crime, trial and execution see L. Worsley, *A Very British Murder: The Story of a National Obsession* (London: Random House, 2013); S. McCorristine, *William Corder and the Red Barn Murder* (London: Palgrave, 2014); for Curtis' role as a crime reporter, see P. Collins, "The Molecatcher's Daughter," *The Independent*, September 26, 2006, accessed May 26, 2015, www.independent.co.uk/arts-entertainment/books/features/the-molecatchers-daughter-425522.html.
7. Curtis, *An Authentic and Faithful History*, 56–7.
8. McCorristine, *William Corder and the Red Barn Murder*.
9. D. Jones, "Thomas Campbell Foster and the Rural Labourer: Incendiarism in East Anglia in the 1840s," *Social History* 1, no. 1: 5–43.
10. Babergh District Council, "Polstead Conservation Area Appraisal," 2012, accessed June 2, 2015, www.babergh.gov.uk/assets/Uploads-BDC/Economy/Heritage/Con-Area Apps/Polstead2012CAA.pdf; Historic England, "List Entry: Maria Marten's cottage," 2015, accessed June 2, 2015, http://list.historicengland.org.uk/resultsingle.aspx?uid=1037040.
11. See P. Bishop, *An Archetypal Constable: National Identity and the Geography of Nostalgia* (London: Athlone, 1995).
12. Worsley, *A Very British Murder*, 91–2.
13. On place-image, see P. C. Adams, *Geographies of Media and Communication* (Chichester: John Wiley, 2009).
14. C. Payne, *Toil and Plenty*; R. Williams, *The Country and the City* (London: Hogarth Press, 1985).
15. M. R. Booth, cited in M. J. Wiener, "Alice Arden to Bill Sykes: Changing Nightmares of Intimate Violence in England, 1558–1869," *Journal of British Studies*, Vol. 4, no. 2 (Apr 2001): 184–212; Kristen Leaver, "Victorian Melodrama and the Performance of Poverty," *Victorian Literature and Culture* 27, no. 2 (1999): 444.
16. Houghton, 1957, cited in Payne, *Toil and Plenty*, 29.
17. See P. Basu, *Narratives in a Landscape: Monuments and Memories of the Sutherland Clearances* (London: University College, 1997); P. Basu, "Cairns in the Landscape," in *Landscapes Beyond Land*, ed. A. Arnason et al. (Oxford: Berghahn, 2012), 116–39.
18. S. Reijnders, "Watching the Detectives: Inside the Guilty Landscapes of Inspector Morse, Baantjer and Wallander," *European Journal of Communication* 24, no. 2 (2009): 175–6.

19 A. Oosterman, "Constructive Guilt," *Volume* 31 (2012): 2–3.
20 E. van Alphen, *Caught by History: Holocaust Effects in Contemporary Art, Literature, and Theory* (Stanford: Stanford University Press, 1997).
21 C. Brants, "Guilty Landscapes: Collective Guilt and International Criminal Law," in *Cosmopolitan Justice and Its Discontents*, ed. C. Bailliet and K. Aas (London: Routledge, 2011), 55.
22 See P. Howard, I. Thompson, and E. Waterton, *The Routledge Companion to Landscape Studies* (London: Routledge, 2005).
23 Council of Europe, European Landscape Convention, 2000, Article 1, accessed 7 March, 2016, http://conventions.coe.int/treaty/en/Treaties/Html/176.htm.
24 J. Assmann, "Communicative and Cultural Memory," in *Cultural Memory Studies: An International and Interdisciplinary Handbook*, ed. A. Erll and A. Nunning (Berlin: de Gruyter, 2008), 109–19.
25 See K. Olwig, "The Law of Landscape and the Landscape of Law: The Things that Matter," in Howard et al., *The Routledge Companion to Landscape Studies*.
26 Assmann, "Communicative and Cultural Memory."
27 A. Rigney, "Plenitude, Scarcity and the Circulation of Cultural Memory," in *Journal of European Studies* 35, no. 1 (2005) 11–35.
28 P. Connerton, "Seven Types of Forgetting," *Memory Studies* 1, no. 1 (2008) 59–71.
29 A. Rigney, "The Dynamics of Remembrance: Texts between Monumentality and Morphing," in *Cultural Memory Studies: An International and Interdisciplinary Handbook*, ed. A. Erll and A. Nunning (Berlin: de Gruyter, 2008), 345–7.
30 See P. Bishop, *An Archetypal Constable*; S. Daniels, *Fields of Vision: Landscape Imagery and National Identity in England and the United States* (Oxford: Polity, 1993).
31 See M. Rosenthal, *Constable: The Painter and His Landscape* (New Haven: Yale University Press, 1983).
32 Dedham Vale AONB and Stour Valley Management Plan 2010–2015, 2010, 8, accessed June 12, 2015, www.dedhamvalestourvalley.org/assets/Publications/Management-Plan-Docs/DV-AONB7996ManagementStrategyPlan.pdf.
33 J. R. Veenstra, "The New Historicism of Stephen Greenblatt," *History and Theory* 34, no. 3 (1995), 174–98; R. Williams, *Culture and Materialism: Selected Essays* (London: Verso, 2005).
34 J. Barrell, *The Dark Side of the Landscape: The Rural Poor in English Painting 1730–1840* (Cambridge: Cambridge University Press, 1980); A. Bermingham, *Landscape and Ideology: The English Rustic Tradition, 1740–1860* (London: Thames and Hudson, 1987).
35 Barrell, *Dark Side*, 133–4.
36 C. R. Leslie, *Memoirs of the Life of John Constable* (London: Phaidon Press, 1951), 85–6; J. Barrell, *Dark Side*, 159.
37 Bermingham, *Landscape and Ideology*, 139.
38 Williams, *Country and the City*, 120.
39 Barrell, *Dark Side*, 157; Bermingham, *Landscape and Ideology*, 142.
40 M. Rosenthal, *Constable*, 191–203.
41 Bermingham, *Landscape and Ideology*, 135.
42 J. Urry, *The Tourist Gaze: Leisure and Travel in Contemporary Societies* (London: Sage, 1990).
43 P. Selman and C. Swanwick, "On the Meaning of Natural Beauty in Landscape Legislation," *Landscape Research* 35, no. 1 (2010): 3–26.
44 M. Johnson, *Ideas of Landscape* (Oxford: Blackwell, 2007); D. Cosgrove, *Social Formation and Symbolic Landscape*, 2nd ed. (Madison: Wisconsin University Press, 1998).
45 L. Taksa, "The Material Culture of an Industrial Artifact," *Historical Archaeology* 34, no. 3 (2005): 8–27.
46 The Stour Valley Landscape Project, "Managing a Masterpiece," accessed June 24, 2015, www.managingamasterpiece.org/

47 The Stour Valley Landscape Partnership, "Managing a Masterpiece," accessed November 16, 2015, www.dedhamvalestourvalley.org/managing-a-masterpiece/
48 Worsley, *A Very British Murder.*
49 C. Hindley, *The History of the Catnach Press* (London: Hindley, 1869), especially xxix, 78.
50 T. Pettit, "Written Composition and (Mem)oral Decomposition: the case of 'The Suffolk Tragedy'," *Oral Tradition* 24, no. 2 (2009): 435.
51 D. Atkinson, "Magical Corpses: Ballads, Intertextuality, and the Discovery of Murder," *Journal of Folklore Research* 36, no. 1 (1999): 1–29.
52 D. Cohen, "The Beautiful Female Murder Victim: Literary Genres and Courtship Practices in the Origins of a Cultural Motif, 1590–1850," *Journal of Social History* 31, no. 2 (1997): 277–306.
53 Pettit, *Written Composition*, 441–2.
54 Leaver, "Victorian Melodrama," 443–56; E. Steele, *Murder and Melodrama: The Red Barn Story on Stage* (College Park: University of Maryland, 2008).
55 A. Briggs, *Victorian Things* (London: Penguin, 1990), 148–50.
56 Foley, 1991, cited in Atkinson, *Magical Corpses*, 16.
57 A letter of Constable from 1821, cited in Rosenthal, *Constable*, 208, 225.
58 E. P. Thompson, *The Making of the English Working Class* (London: Victor Gollancz, 1963); E. Hobsbawm and G. Rude, *Captain Swing* (London: Verso [1969] 2014); B. Reay, *The Last Rising of the Agricultural Labourers: Rural Life and Protest in Nineteenth-century England* (Oxford: Oxford University Press, 1990).
59 M. Rosenthal, "This Green Unpleasant Land: Landscape and Contemporary Britain," in *The Place of Landscape: Concepts, Contexts, Studies*, ed. J. Malpas (Cambridge, MA: MIT Press, 2011), 273–95.
60 L. Lippard, *The Lure of the Local: Senses of Place in a Multicentered Society* (New York: New Press, 1997), 8.
61 A. Oosterman, "Constructive Guilt," 3.
62 E. Wyschogrod, *An Ethics of Remembering: History, Heterology and the Nameless Others* (Chicago: University of Chicago Press, 1998).

7 A culturally constructed darkness
Dark legacies and dark heritage in the Channel Islands

Gilly Carr

Legacies and heritage

As a result of the German occupation of the Channel Islands from 1940 to 1945, many places contain potential residues of 'darkness': the sites of forced and slave labour camps, the prisons, the places of deportation and the bunkers. But how does the local population cope with such a great and undeniable 'darkness' within their small islands? This chapter addresses the question of where the darkness originates. It argues that the darkness of this legacy is not fixed; neither does it emanate from the sites themselves, even for people who know and understand the meaning and history of such places. Rather, it suggests that it is in the eye of the beholder, perceived differently by locals and visitors, and is indeed culturally determined and understood.

The Channel Islands were occupied by German forces at the end of June and beginning of July 1940; they were the last places in occupied Europe to be liberated, on 9 May 1945, a national holiday in the islands today. Although the Channel Islands were the most leniently treated of all German-occupied places in Europe, this is not to say that dark things did not happen in the islands or that, today, they are not full of potentially dark legacies. We might imagine, for example, that the islands' harbours were among the darkest of locations as they were the scenes of forced deportation and importation of people. Islanders who committed acts of resistance were imprisoned locally or deported to Nazi-controlled prisons or concentration camps on the continent.[1] The islands' tiny Jewish communities were persecuted and deported,[2] and English-born islanders were among those deported to civilian internment camps in Germany in September 1942 and February 1943.[3] The harbours were also the sites of arrival for thousands of forced and slave labourers from all over Europe to build the bunkers of the Atlantic Wall, many of whom arrived in a shocking condition, were treated appallingly and were housed in labour camps in the islands.[4]

The harbours, which formed the backdrop to many forced partings and arrivals during the occupation, were in reality but places of transition. No 'dark aura' appears to adhere to them today. However, this statement needs examination before it can be so easily dismissed. The harbours of all of the Channel Islands are places where a multiplicity of memorials to the years of occupation has mushroomed over the past fifteen to twenty years.[5] Why would this be necessary if these sites were not dark – even if a lighter shade of darkness on Stone's spectrum[6] – to some degree? For Stone, 'lighter' sites were those 'associated with' death and suffering, while those that were 'darker' were those 'of' death and suffering. But for whom are they dark at all? For sure, the harbours are not merely random or convenient placements for such memorials; they are genuine *lieux de mémoire*.[7] The more memorials that have been erected, the more the harbours become seen as the *proper* place for them, regardless of whether they are always the most *appropriate* place. Memorials themselves can have their own emotional or dark impact through their choice of text, especially for the people or family members of the group they represent.

In this chapter, however, I argue that not only have memorials been placed in convenient and substitute non-dark *lieux de mémoire*; places which are genuine sites of memory but which nonetheless are only transitional 'places on the journey' – the 'moveable feasts' and preludes to the main act. However, there exist other sites of memory, such as labour camps and prisons. These have become *lieux d'oubli*, and it is here where the darkness really resides, or has the potential to reside if uncovered or recognised as heritage. These forgotten sites or 'sites of oblivion' are, as Nancy Wood[8] explains, intentionally avoided by public memory 'because of the disturbing affect that their invocation is still capable of arousing'. Before we can explore these forgotten dark sites, which are not recognised as public heritage sites, it is important to characterise them in terms of their state of existence or becoming. Only then can we understand which sites are avoided and which draw public attention and why: why darkness is perceived to adhere to one type of site and not another, and why this perception may differ between tourists and locals. In order to address this, I would like to propose a schematic model to show the relationship between a dark event, a dark legacy, dark heritage and dark tourism, as shown in Figure 7.1.

A dark *event* (such as military occupation) leaves behind it a tangible *legacy* or residue in the form of traces, ruins, debris, sites and objects, and an intangible legacy in the form of memories, trauma and psychological impact. All of these have the potential to be dark. If we take the example of dark sites, which might take the form of prisons, labour camps or military fortifications, none of these automatically deserve the status of 'heritage'. Using as my point of departure Smith's concept of heritage, which she defines as 'the cultural processes of meaning and memory making and remaking rather than a thing',[9] I suggest that heritage has to be created or chosen through an active process. For example, if a fortification is left untouched after a dark event, then sooner or later it will succumb to the passage of time. It is still at the *legacy* stage with no certainty that it might ever become anything else. If not maintained, the fabric of the building will start to degrade and even collapse. This is not heritage; *heritage* is something (tangible or intangible) that is valued, selected and chosen to represent some aspect of identity of a group. For Channel Islanders, bunkers have been reclaimed from the weeds and from oblivion since the late 1970s. At and after this point, many have been turned into museums or restored to how they would have looked when operational. This active intervention has turned them into heritage. But heritage can just as

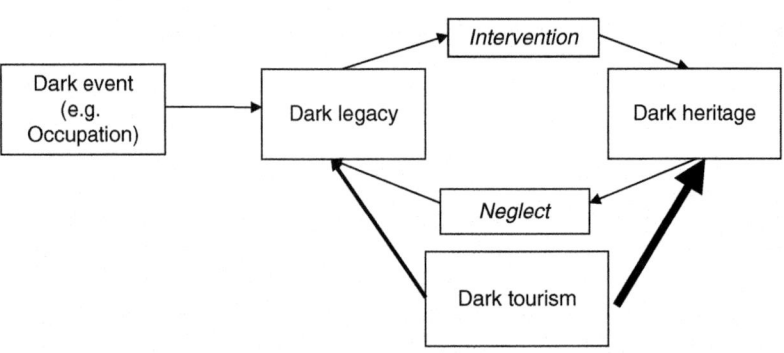

Figure 7.1 The links between dark events, legacies, heritage and tourism.

easily be abandoned and neglected, and revert to legacy status. Thus, the passage between legacy and heritage can be cyclical and not linear. When tourists visit sites of German occupation in the Channel Islands, they are arguably more likely to visit heritage sites, where there is 'something to see'; more likely (in this example) to visit bunker museums or restored bunkers rather than abandoned bunkers hidden in the undergrowth. In fact, such legacies are less known about by tourists and thus, predictably, more tourists visit heritage than legacy sites (represented by the different arrow thicknesses in the diagram).

This much may be true for tourists, but it is perhaps not so true for non-tourists. In the Channel Islands, a popular pastime at weekends is to 'go bunkering'. This involves visiting neglected, hidden or out-of-the-way bunkers still at 'legacy' status. It is here, and not in the restored bunkers, where the 'darkness' may be perceived locally, as I will explore later. But what is the real difference between bunkers of legacy and heritage status? One might imagine that heritage-status bunkers have been more thoroughly researched and explored, and thus the greater knowledge of the human rights abuses against slave and forced workers that took place there would lead to a heritage which proclaimed that history and acknowledged that 'darkness'. One might also imagine that the people who have spent many years restoring and maintaining the bunkers might be those whose intimate knowledge of the sites has resulted in a familiarity and awareness, and who embrace its 'dark' past. But this is not the case; these kinds of bunkers are not perceived locally to be 'dark' places. We might assume that long familiarity has bred a contempt or lack of respect for the darkness emanating from these sites of death and suffering. But such an assumption would be to ignore the impact of cultural identity, and it is cultural identity rather than historical knowledge *per se* that dictates not only from where darkness emanates, but also who can perceive it in such places.

It is a curious fact that, while the Channel Islands had a similar experience of war to that of other occupied countries in Western Europe (including the persecution of Jews, the deportation of people who committed acts of resistance, the importation of slave labourers, starvation and hunger, etc), their war narrative is far removed from that of other European countries. Rather than perceiving themselves, as other Europeans do, as victims of occupation, or as martyrs for having been sent to their deaths in prisons or camps for standing up to the Nazis, the Channel Islands have adopted the British war narrative. As Britons who were liberated by British military forces, they shared in a British victory and thus saw, and see, themselves through a Churchillian lens. According to Paul Sanders,[10] the 'Churchillian paradigm' embraced a narrative of 'sublime and unwavering steadfastness in the face of adversity' as soon as the war came to an end. This paradigm states that 'the British were not a nation of victims, but of victors'[11] – a narrative of heroic victory which excludes any rival version of events and disregards the multiple and divergent memories of occupation. However, rather than using bunkers as blank canvases for telling the story of an allied victory and the post-war use of bunkers as the reclamation of Channel Island territory, the Churchillian paradigm instead dictates that bunkers are places to emphasise the strength and power of the enemy – an enemy that was overthrown. Thus, the display of guns and other militaria, uniforms, helmets and swastikas show off the spoils of war, the booty that became the property of the victor. Bunkers therefore did not become places to tell the story of victimhood, the narrative of the slave or forced labourer, dragged across Europe to build bunkers, worked or starved to death; instead, they became 'dark-proof', where the only admitted narrative was one of the defeat of mighty Goliath, with the Channel Islands adopting the role of David. Thus, it is only restored bunkers – bunkers-as-heritage – that have been converted to this use and which proclaim the dominant narrative, while bunkers-as-legacy are perceived as telling a very different story. Such a dominant narrative can also be seen in terms of what Bell[12] calls a 'governing myth', where a 'mythscape' is the 'temporally and spatially extended discursive realm wherein the struggle for control of people's memories and the formation of nationalist myths is debated, contested and subverted incessantly'.[13] While the majority of restored bunkers and bunker museums in the Channel Islands are places where the Churchillian paradigm as governing myth is writ large, there are a small number – just two – which have been used for other purposes, and which reveal the space within bunkers as locations of mythscapes.

In Jersey, at La Hougue Bie and at Jersey War Tunnels, are two German fortifications whose original heritage presentation espoused the Churchillian paradigm and which are now, after conversion, upgrading and reconceptualisation by heritage professionals rather than amateur enthusiasts, places which embrace the role of bunkers as places where people suffered. The foreign labourers are given back their names and identities and their individual stories are told. This contestation and subversion in the 'governing myth' of occupation has been controversial. Bunkers have become symbolic space where whoever owns or controls that space also controls the narrative within it. Indeed, the 'darkness' in both of these renegotiated bunker sites has now begun to be acknowledged. I argue, therefore, that the 'darkness' lies not entirely within the 'auratic quality' which such a site might emit, nor in the knowledge of what it was used for, but in the dominant narrative that prevails in a place, and which often dictates the associated heritage interpretation. I also argue that local people are more likely than tourists to be affected by, or wholly cognisant of, the governing myth, having grown up with it, been taught it in school and seen it enacted in the streets on both ceremonial and celebratory occasions in the local calendar, such as on Liberation Day. Visitors and tourists, on the other hand, will be less 'indoctrinated' into or affected by governing myths. They may be ignorant of the governing myth and unaware of such uninterpreted legacy-status sites, and head only to the heritage sites. They may perceive darkness to emanate, in other words, from the very sites that local people do not perceive as dark.

Let us now focus on these so-called legacy/non-heritage sites, among which I include labour camps, unrestored bunkers and prisons. While their 'darkness' might be perceived by local people to stem from their potential status as 'terrorscapes',[14] this 'darkness' is arguably retained in these sites and denied by local people to exist in heritage sites, such as restored bunkers and places of deportation in the harbours. Because 'darkness' is something that is in the eye of the beholder, it exists in different places for different people. The sites of the local wartime prisons in St Helier and St Peter Port (the capital towns of Jersey and Guernsey respctively) have long been demolished and have not been kept as sites of martyrdom and victimhood. Such sites would be contrary to the British war narrative of victory, which stands in opposition to such experiences. Embraced in the Channel Islands since the earliest post-war days, this narrative proclaims the myth of 'correct relations' with the occupiers, 'devoid of the humiliation, desperation or compromise of principle occurring across the rest of Europe'.[15] However, it is the *lieux de memoire* (or, rather, *lieux d'oubli*) of these other groups, the victims of Nazism, which have been avoided, often destroyed, and certainly not turned into heritage. Existing primarily only as a legacy or a memory, these are the places which have the power to disturb and intrude on public memory. Where they have been turned into heritage in recent years because of a challenge to the status quo from outsiders or family, or even members of victim groups, the sites have become places of counter-memory. As Legg suggests, these are the places which 'mark times and places which people have refused to forget. They can rebut the memory schema of a dominant class, race, or nation, providing an alternative form of remembering and identity.' These are the sites where people have challenged dominant narratives and have nurtured the survival of memories beyond the memory sites of the nation.[16]

Although Channel Island harbours are but transitional places on the journey to where the real sites of memory were for many, they were also stepping stones en route to prisons and camps where imported labourers were accommodated and brutalised. There were around twelve slave and forced labour camps in Jersey and around five each in Guernsey and Alderney, not including the reuse of houses and other buildings in the islands for accommodation and penal prisons of the labour force. Alderney was also the location of a concentration camp run by the SS, *Lager* Sylt.

The manual labourers of the Organisation Todt (OT), a paramilitary engineering organisation, were a workforce which reached a height of around 16,000 in May 1943.[17] They comprised voluntary, conscripted and forced workers, but also slave labourers. Among these were heterogeneous groups that included Jews who had been rounded up in occupied Europe, but also Poles, Czechs, Belarusians, Ukrainians, Russians, North Africans and Spanish Republicans, as well as German 'criminals' and political prisoners.[18] Even local Channel Islanders, attracted by the high rates of pay, worked for the OT as cooks, interpreters, drivers and skilled labour.[19] While various nationalities and groups of the labour force were treated better than others, and some even paid and given time off, the Russians, Ukrainians and people considered by the Nazis to be of 'Slavic origin' were categorised as *Untermenschen* or sub-human, treated appallingly and often brutally and given very little food. The number of deaths of slave workers in the islands is disputed and will probably never be known with any accuracy, but figures range from 389 to 5,000 people.[20] These sources and others also record the testimony of a number of former slave workers, and the extreme suffering of these people is beyond doubt. Although a summary of this is outside the scope of this chapter, it will suffice to record that the unremitting hard labour involved, coupled with the starvation rations, poor living and sanitary conditions, overseer violence and lack of medical attention, directly contributed to or caused most of the deaths.

Walter argues that one of the factors which can conspire to darken sites of death is those deaths which 'challenge the collective narratives of a nation'.[21] We have already noted that the war narrative of the Channel Islands has traditionally and specifically excluded victims, especially victims of Nazism, and therefore one of the reasons why these camps have been obliterated or ignored is because of their power to upset the heritage status quo and governing myth. It is true to say that not a single one of these camps today has become a heritage site. Every single one of them is at the status of a legacy or just a memory, and most are completely destroyed and have modern buildings built upon them. Even this alone is an interesting observation, for who would want to live on top of a labour camp? It is unknown how many people know the wartime history of the patch of ground in the islands upon which they live, and it is possible that the passage of time since the war has lessened the power or perception of the darkness which might adhere to such sites. In Alderney, for example, there are bungalows built on top of *Lager* Helgoland, a labour camp which was thought to contain Jews.[22] Here, the former entrance posts to the camp are used as the entrance posts to the driveway of the bungalow. At *Lager* Norderney, now the island's camp site for tents and caravans in summer, various concrete structures are still visible in the long grass. Local people, in such a tiny island, are surely aware of what they live and camp upon. The traces of *Lager* Borkum in Alderney today stand either side of the track that leads to the island's rubbish dump. The ruins of *Lager* Sylt lie abandoned, covered by the lush vegetation for which the island is famed. In 2008, this camp became the first and only one in the islands to receive a memorial plaque, but even this was placed on the entrance post of the camp by a former Polish prisoner and was not a local initiative. More than in Jersey and Guernsey, the camps are sites of oblivion in Alderney. There has even been a marked reluctance to support archaeological excavation in Alderney[23] – clear evidence of these camps' status as *lieux d'oubli*.

The situation is little better in Jersey and Guernsey, although fewer traces remain. Rather than being sites which people expressly avoid, more of them have been obliterated in the landscape. It seems that while traces remain, the power of the darkness of these sites is too much to tolerate. While labour camps have been categorised as merely the 'accommodation' of the men who worked on the bunkers,[24] a neutral word that implies nothing sinister, we know that reports exist in Jersey of Russians being kept in cages or wire compounds within their camps, such as at *Lager* Udet in St Brelade, and some camp commandants were known for their brutality, such as at *Lager* Immelmann in the parish of St Peter.[25] There are also plenty of reports by local people of foreign workers' torture and bad treatment by their overseers. One of the better known accounts was that of Senator Edward Le Quesne in Jersey, whose diary entry for 20 February 1943 recorded that he had seen in the parish of St Ouen a Russian in the pillory with two branches of trees tied tightly around his neck, the man just able to touch the ground with his toes. As he had an armed guard standing over him, nobody was able to help him.[26]

While one might have imagined the labour camps of the Channel Islands to be the prime sites of darkness today, this is apparently not the case, because they are simply ignored. It is possible that these places have simply lost much of their darkness because they have been neutered through destruction or neglect. But by not confronting what happened at these sites, or to the people who were forced to reside there, local people are able to continue ignoring them. This does not mean that these sites do not have the power to grow darker if ever they are uncovered and draw an audience. To excavate sites such as these, for example, is to open Pandora's box and risks revealing something which people may not yet be prepared to face. It is known, for example, that Jews were among those brought to the Channel Islands to work for the OT, and this adds to the potential feelings of anxiety about what could yet be revealed.[27]

While OT camps are not presented as tourist sites, it is not entirely true to say that they are not visited. Local historians or researchers sometimes visit these sites, and former OT workers have also made the pilgrimage back to the sites of their suffering. Photographic evidence exists of this in Jersey from around 1970, when resident Spanish Republicans visited the sites of camps in the island.[28] A similar event also took place when the memorial plaque was attached to the gate post of the concentration camp of Sylt in Alderney in 2008. It is difficult to tell whether the general dissipation of the darkness of OT camps in Guernsey and, especially, Jersey happened slowly over several generations or, as is more likely, whether the real neutering of the camps took place quite early on, when they were destroyed or dismantled by the Germans and locals alike. It is not known whether the motivation for the destruction or removal of any of the camps stemmed from a desire to cover up the evidence of their crimes (in the case of the Germans), or not to be reminded of the crimes that took place on their soil (in the case of the islanders), or a combination of both. While we may wonder at islanders' lack of anxiety over their islands' role in the Holocaust, we should remember that it was not until the early 1970s, after the Eichmann trial of 1961, the 1967 Arab–Israeli war and the 1968 student protests, that the Holocaust began to assume centre stage in the consciousness of Western Europeans.[29] By this time, the state of preservation of most OT camp sites in the Channel Islands may not have been too different to their status today, as the photos taken around this time by Spanish Republicans in the island attest.

Should we be concerned that most of these sites have apparently lost their power to disturb? Is it sometimes a good thing for dark sites to lose their darkness? On the one hand, it means that communities can reclaim the land for the living, let go of the past and move forward, ridding themselves of the burden of war, all of which might be perceived as a thoroughly healthy and positive thing several generations after the original conflict. On the other hand, there are ethical ramifications to ignoring such sites. Even the apparently innocuous camps of many forced labourers from Western Europe (as opposed to slave labourers from Eastern Europe) housed people who were taken against their will or who had little choice but to agree to work for the OT rather than face an unknown fate in Germany.[30] Human rights abuses within a corrupt system were endemic inside the OT, and to differentiate between camp types, or to label some as darker than others depending on who lived there and how they were treated, is to turn one's back on past suffering. However, the current neglect of OT camps does not necessarily indicate their terminal position. The legacy of war does not have a pre-ordained trajectory or life-cycle; something that is covered in the undergrowth or long forgotten does not have to stay in this state. Interventions by archaeologists or any other stakeholders to uncover and preserve the camps are possible, but the success of these efforts will be dictated either by the local community or by those in positions of authority who have the power to sanction or loudly welcome such interventions. However, as Geyer wisely points out, 'no preservation, however perfect, can save these traces for the present unless they are accepted in the present'.[31] An uncovered site imposed upon the local community as 'heritage' can return to its previously neglected state if locally rejected or disliked. The converse is also true.

But if members of the local community show no sign of wanting to change the status quo, as seems apparent in the Channel Islands, do outsiders have any right to intervene and engineer or impose a change of any kind – to force them to face the darkness? If local people are not capable of discerning the darkness at a site, or deny that the darkness exists, can and should darkness be forced upon them for honourable or ethical reasons, such as raising awareness about the people who once suffered there, or for reasons of education, or to try to change the war narrative of a place through force? How possible or ethical is such an attempt? Is a desire to show respect to victims of Nazism enough to claim the moral high ground? We also must not lose sight of *why* members of the local community have not turned the camps into heritage. Such a decision makes a statement about what Channel Island identity does and does not embrace, with the associated implication that to force a change is an attempt to manipulate or misrepresent locally held concepts of identity and even collective memory. Arguably, the decision to ignore this legacy of occupation was taken many decades ago, and the subject has never re-arisen for debate. Since the fiftieth anniversary of liberation, however, islanders are more open to embracing and remembering victims of Nazism, especially in Jersey; progress has been made more slowly in Guernsey and is hardly detectable in Alderney. However, if the subject of camps were discussed again today, it is possible that the outcome could be different. With this in mind, in 2014 and 2015 I began the very first excavation of a labour camp in the Channel Islands. I carried out work at *Lager* Wick in

Jersey, a forced labour camp for men from France, Spanish Republicans and French North Africans,[32] today a nature reserve for wild birds. The excavation blog was followed by people from fifty-five and forty-eight different countries in 2014 and 2015 respectively, the highest number of 'hits' coming, perhaps not surprisingly, from the UK and Jersey. While the excavation also attracted coverage from the local Channel Islands media, I discovered that the labour camp I had excavated was either not perceived as 'dark' by local people, or else any unmistakably dark elements were either being denied permanent exposure or given a lighter spin in accordance with the Churchillian paradigm and its avoidance of a narrative of victimhood. Three examples will suffice to back up this observation.

The only remaining features of *Lager* Wick above ground were its concrete entrance posts, complete with several strands of barbed wire wrapped around them (Figure 7.2). These had previously been covered in so much ivy that the

Figure 7.2 The entrance posts of *Lager* Wick, Jersey.
Photo copyright G. Carr.

A culturally constructed darkness 105

posts appeared entirely indistinguishable from the surrounding trees. During the first season of excavation I stripped back the ivy, revealing the posts once again for the first time in decades. Such structures were undoubtedly, to my eyes, almost iconic features of a Nazi camp, and my recommendation to the land owners and the local planning authorities was to leave them uncovered as part of presenting the site to the public as a heritage site. This recommendation was turned down as it was deemed more important for the wildlife to restore their habitat to the state that it had been in before I arrived, thus covering up once again the only recognisable dark feature of the camp.

A few months after the first season of excavation, I was sent a PDF of an artist's impression of the camp and some text which would be placed on an information board by the side of the road by the camp. I was rather surprised to see a sanitised image of a spotlessly clean and orderly series of barrack huts with no hint of squalor or barbed wire, and the concrete entrance posts had been omitted. Representations to those who manage the site resulted in slight changes, but still no sign of barbed wire or the entrance posts graced the final image which is now in place outside the site (Figure 7.3).

During the second season of excavation I uncovered both the ablutions block of the camp and a barrack block which, by the end of the excavation, I interpreted as belonging to the camp overseers because of the nature of the objects discovered. The excavation of the ablutions block made the lofty heights of page three of the

Figure 7.3 Information panel about *Lager* Wick, Jersey.
Photo copyright G. Carr.

Jersey Evening Post, although the article focused primarily on my call for local volunteers for the dig rather than on our discoveries. Then, on the penultimate day, I discovered a mug which featured an eagle and swastika on the base, which made the front pages the following day. I imagined that this might be an opportunity to discuss the role of the overseers and their ill treatment of the camp inmates, but, after printing my quote saying that it was time to acknowledge the darker side of the island's heritage, the focus instead was on how 'Dr Carr [had] secured agreement that the ablutions block would be preserved as a heritage site'.[33]

Conclusion

In this chapter I have suggested that the darkness of a place is neither fixed nor given, and nor does it emanate from the site as an 'aura'. Rather, it is in the eye of the beholder and is primarily culturally constructed and understood, able to be denied, destroyed or marginalised by those who do not wish to see or feel it. This darkness, I have argued, comes from the ghosts of the past which haunt the ignored legacies of occupation. In the Channel Islands the Churchillian paradigm – with its avoidance of victims of Nazism and its focus on victory, and a 'governing myth' of correct relations with the Germans – has led to the avoidance and denial of the dark. Islanders are, on some level, haunted by these ghosts. They are aware of them but will draw instead upon less dark and less traumatic narratives of the past, even going to the extent of destroying structures which had the power to betray dark residues. While the casual tourist to the Channel Islands will see dark heritage everywhere, islanders themselves do not see heritage that way. Sites which have been chosen by them as heritage have been selected because they are not dark to their eyes, and do not tell a story of darkness; rather, they are made to conform to the governing myth, which is part of Channel Islands cultural identity. The dark *legacies* of occupation, on the other hand, are where the real darkness lies for them, and these have been destroyed, marginalised or ignored as sites of oblivion. Even when confronted with potential darkness, the eyes of these beholders paint it in lighter colours.

Notes

1. G. Carr, P. Sanders, and L. Willmott, *Protest, Defiance and Resistance in the Channel Islands: German Occupation 1940–1945* (London: Bloomsbury, 2014).
2. F. Cohen, *The Jews in the Channel Islands during German Occupation, 1940–1945* (London: Jersey Heritage Trust/The Institute of Contemporary History and Wiener Library Ltd, 2000).
3. R. Harris, *Islanders Deported* (Ilford: CISS Publishing, 1979); G. Carr, *Occupied Behind Barbed Wire* (Jersey: Jersey Heritage, 2010).
4. M. Ginns, *The Organisation Todt and the Fortress Engineers in the Channel Islands* (Jersey: Channel Islands Occupation Society Archive Book No.8, 2006).
5. G. Carr, "Examining the Memorialscape of Occupation and Liberation: A Case Study from the Channel Islands," *International Journal of Heritage Studies* 18, no. 2 (2012): 174–93.
6. P. Stone, "A Dark Tourism Spectrum: Towards a Typology of Death and Macabre Related Tourist Sites, Attractions and Exhibitions," *Tourism* 54, no. 2 (2006): 145–60.

7 P. Nora, "Between Memory and History: Les lieux de mémoire," *Representations* 26 (1989): 7–24.
 8 N. Wood, *Vectors of Memory: Legacies of Trauma in Postwar Europe* (Oxford: Berg, 1999), 10.
 9 L. Smith, *Uses of Heritage* (London: Routledge, 2006), 74.
10 P. Sanders, "Narratives of Britishness: UK War Memory and Channel Islands Occupation Memory," in *Islands and Britishness: A Global Perspective*, ed. J. Matthews and D. Travers (Newcastle: Cambridge Scholars, 2012), 25.
11 P. Sanders, *The British Channel Islands under German Occupation, 1940–1945* (Jersey: Société Jersiaise and Jersey Heritage Trust, 2005), 256.
12 D. Bell, "Mythscapes: Memory, Mythology, and National Identity," *British Journal of Sociology* 54, no. 1 (2003): 63–81.
13 Ibid., 66.
14 R. van der Laarse, "Beyond Auschwitz? Europe's Terrorscapes in the Age of Postmemory," in *Memory and Postwar Memorials: Confronting the Violence of the Past*, ed. M. Silberman and F. Vatan (New York: Palgrave, 2013), 71–92.
15 Sanders, *The British Channel Islands under Occupation*, 235.
16 S. Legg, "Sites of Counter-memory: The Refusal to Forget and the Nationalist Struggle in Colonial Delhi," *Historical Geography* 33 (2005): 181.
17 C. Cruickshank, *The German Occupation of the Channel Islands* (London: Oxford University Press/Channel Islands: Guernsey Press, 2004 [1975]), 204.
18 Cohen, *The Jews in the Channel Islands during the German Occupation*, 122, 130.
19 Ginns, *The Organisation Todt and the Fortress Engineers in the Channel Islands*, 64–7; M. Bunting, *The Model Occupation* (London: BCA 1995), 94–5.
20 See Bunting, *The Model Occupation*, 293; Cruickshank, *The German Occupation*, 213–14; Cohen, *The Jews in the Channel Islands*, 147–52; Ginns, *The Organisation Todt*, 115–25; H. Knowles Smith, *The Changing Face of the Channel Islands Occupation: Record, Memory and Myth* (Basingstoke: Palgrave Macmillan, 2007), 209–30; T. Pantcheff, *Alderney: Fortress Island* (Sussex: Phillimore, 1981); Sanders, *The British Channel Islands under German Occupation*.
21 T. Walter, "Dark Tourism: Mediating between the Dead and the Living," in *The Darker Side of Travel: The Theory and Practice of Dark Tourism*, ed. Richard Sharpley and Philip R. Stone (Bristol: Channel View, 2009), 52.
22 Ginns, *The Organisation Todt*, 85.
23 C. Sturdy Colls, "Holocaust Archaeology: Archaeological Approaches to Landscapes of Nazi Genocide and Persecution," *Journal of Conflict Archaeology* 7, no. 2 (2012): 94.
24 Ginns, *The Organisation Todt*, 74.
25 Ibid., 78–80.
26 E. Le Quesne, *The Occupation of Jersey Day by Day* (Jersey: La Haule, 1999).
27 See F. Cohen, *Jews in the Channel Islands*, 121–54; Sanders, *The British Channel Islands under German Occupation*, chapter 6.
28 Gary Font, personal communication.
29 C. Koonz, "Between Memory and Oblivion: Concentration Camps in German Memory," in *Commemorations: The Politics of National Identity*, ed. J. R. Gillis (New Jersey: Princeton University Press), 269.
30 Sanders, *The British Channel Islands under German Occupation*, 205.
31 M. Geyer and M. Latham, "The Place of the Second World War in German Memory and History," *New German Critique* 71 (1997): 7.
32 Ginns, *The Organisation Todt*, 74.
33 P. Thelwell, "Labour Camp Remains 'Should Become Occupation Memorial,'" *Jersey Evening Post*, April 3, 2015, 8.

8 A light in dark places?
Analysing the impact of dark tourism experiences on everyday life

Ria Dunkley

In recent years, tourism scholars have urged the use of more critical methodologies, to the extent that an alternative paradigm of so-called 'hopeful tourism' has emerged.[1] In tandem, a rejection of hegemonic forms of travel has led to the development of arguably 'critical' or 'hopeful' tourism niches, including graffiti, slum and counter-tourism initiatives, as well as urban exploration and the phenomenon discussed within this chapter: thanatourism, or dark tourism, as it is more commonly known. Dark tourism initiatives frequently profess a social mission, with site providers arguing that the interpretation of dissonant heritage promotes peace through tourism by way of, for example, (re)educating visitors. But to what extent can and do such sites present their attendees with opportunities to critically reflect on their everyday lives, to a degree that could result in the cultural shift required to end world injustices? The answer, to some extent, lies in ephemeral moments of transformation. The attitudinal and behavioural shifts brought about through such moments of transformation are best understood by in-depth analysis of the dark tourism experience, gained via interpretive qualitative social scientific study. This chapter presents empirical findings of such a study of tourist experiences at dark tourism sites that include Auschwitz-Birkenau, the battlefields of the First World War and a murder tour of London. Findings demonstrate the lasting impacts and continuing deeper resonance that visits to dark tourism sites can have. Yet dark tourism's capacity to lead visitors from empathy towards compassion, an arguable prerequisite for positive action within the current day, is called into question. Such findings further our understanding of dark tourism as a field, but have implications for how such sites function socially and politically in helping to shape attitudes and beliefs.

It is often claimed that the visitation of death sites is beneficial to societal progress. Dark tourism is thought to provide individuals with the opportunity for deep reflection upon human suffering, and in this respect, the importance of remembering past atrocities appears to be an accepted good.[2] European sites such as Auschwitz and the battlefields of the First and Second World Wars – as well as those in south-east Asia, including the S-21 prison in Cambodia and the Cu Chi Tunnels in Vietnam – are thought to stand testament to the horrors that occurred within the spaces they occupy.[3] As well as having a commemorative role, they are thought to contribute to ensuring that tragic events do not reoccur. There is perhaps no stronger argument than to use Holocaust survivor Primo Levi's words relating to the continued existence of Auschwitz:[4] as Levi put it, the events of the Holocaust 'are not mistakes to efface. With the passing of years and decades, their remains do not lose any of their significance as a warning monument; rather, they gain in meaning.' Further, it has been suggested that destroying such sites would assist perpetrators in masking their evil deeds.[5]

In seeking to make an initial contribution to this debate, this chapter addresses the question of whether visiting dark tourism sites enables critical reflection on everyday life. It does so by drawing upon the results of an in-depth study of dark tourist motivations and experiences. Following the interpretive tradition, insights were gained through creative conversations with dark tourists, through which their individual narratives of experiences emerged.[6] By providing such insights, the chapter will explore the question of whether experiencing dark tourism appears to influence individual tourists' ways of seeing. In achieving this, it is possible to go some way towards inferring whether dark tourism experiences could contribute to a culture shift that would demonstrate the phenomenon's capacity to be a vehicle for peace through tourism. The chapter will explore the existential nature of the dark tourism experience from the tourist's perspective. It provides insights into the meaning of such experiences within the lives of those who become dark tourists, albeit often temporarily.

Gaining a wider understanding of the dark tourism experience is an important development within the field, enabling us to bring to light its wider social and cultural roles. This would allow us to appreciate whether dark tourism is more than a niche market within a burgeoning heritage industry. Yet as well as seeking to deepen insights into the dark tourism experience, the chapter also problematises claims made about such sites, opening up the rarer discussion concerning whether dark tourism can have a negative impact upon individuals and communities. The chapter seeks to explore what is really happening for an individual when they visit a dark tourism site. It concludes that while dark tourism sites are very likely to unveil feelings of empathy, dark tourists less commonly discuss feelings of compassion. This is essential given that compassion, unlike empathy, is likely to be an emotional response that acts as a motivator for action. The chapter therefore summates that for dark tourism sites to be truly transformational there is a greater need to focus upon nurturing compassion through dark tourism, rather than the empathetic feelings that seem to come from reflection on the past alone.

Dark spaces: contributing to peace through tourism?

Although the existence of dark tourism sites is not a new phenomenon, it could be argued that the urge to commemorate has increased in recent times. This is true at least within the Western world, where more memorial museums have opened in the early twenty-first century than was the case in all of the twentieth century collectively.[7] It is suggested that such commemoration is beneficial, given that dark events are part of the mass conscience, and thus it is important that societies remember and progress beyond them.[8] Dark tourism sites are seen as significant, especially in their capacity to raise awareness among 'future generations' of young people who have no lived experience of the past tragedies associated with them. There are few signs of sites being perceived as of decreasing importance, while it could be suggested that greater emphasis is being placed on visiting death sites to learn from the past. For example, in 2014, the UK government established a multi-million pound educational initiative that enabled two students and one teacher from every state school in England to visit the Western Front battlefields as part of the First World War centenary commemorations.[9] The First World War Centenary Battlefield Tours Programme will run until 2019 and provides free tours of the Somme Battlefields. The aim of these visits is to provide young people with a 'personal connection to the First World War'.[10]

As well as seeking to commemorate past tragedies, dark tourism sites are often explicit in their moral mission of ending human suffering. In this respect, they join a host of other initiatives, including peace parks, community tourism and eco-tourism initiatives, in seeking to fulfil the global objective of peace through tourism.[11] To greater and lesser degrees, dark sites make the claim that their presence makes repetition of tragic events less likely. Warfare tourism sites in particular often stress their ability to contribute towards global peace. To this end, Uzzell states that 'If handled with care and sensitivity atrocity and war sites can play a vital role in improving understanding and raising awareness of human cruelty and suffering and perhaps help to avoid its repetition'.[12] It could also be argued that sites of hardship – for example, notable prisons, including Robben Island – may have an important role to play. Furthermore, sites associated with tragedies – including the Hiroshima Peace Memorial Park – as well as sites of the murders of notable individuals – including, for example, the Strawberry Fields Conservancy, New York and the Sixth Floor Development, Dallas – also make a case for their role within societal progress.

A key consideration from a social science perspective may be whether such dark tourist experiences contribute towards pro-social goals in the sense that dark site proprietors often argue is possible. Much criticism is aimed at dark tourism sites and indeed dark tourists. It is often suggested, for instance, that this form of tourism is a commodification of tragic past events by a tourism industry which serves to cater for the morbidly curious tourist. It has been reported that sites may incite negative emotions such as anger, which may be provoked by a sense of injustice felt by site visitors who may as a result still consider themselves to be on an opposing side to other site visitors. This was found to be the case at the Pearl Harbor memorial in Japan, where tourists left angry graffiti.[13] In this respect, dark tourism may thus serve to heighten tensions between peoples who, because of narratives presented on site, may be encouraged to regard each other as enemies. Finally, despite claims that the existence of dark tourism may prevent future conflicts and tragedies, tragic events continue to occur globally, albeit in differing contexts. As Rupesinghe puts it, despite the preservation of Holocaust sites and the cinematic interpretation of its horror, 'as cinema audiences recoiled from the gruesome scenes in a Hollywood movie, the Muslim population of Bosnia were reliving them'.[14] This may lead some to conclude that those who perhaps would benefit the most from visiting such dark sites are either not doing so or remain unaffected by their experiences should they have done so.

Yet, much of this speculation concerning the need to interpret tragic pasts for a touristic audience remains unsubstantiated. Whether dark tourism sites, including memorial museums, are effective in their commemorative capacity or in imparting values is seldom critiqued beyond a superficial level of consideration.[15] Despite such assertions from the industry, from the academy, and indeed from policy-makers concerning the importance of dark tourism sites within contemporary society, very little is known about the role of dark tourism experiences within the contexts of the everyday lives of the individuals who choose to visit them. Currently the dark tourism literature, as an emerging field, is yet to present an in-depth understanding of whether dark tourism has a profound impact upon tourists to the extent that visiting dark tourism sites would influence understandings and behaviours in order to make them capable of contributing towards peace. Such contributions might include, for example, the ability to nurture greater intergenerational and intercultural understanding, increased compassion and motivation to participate in democratic processes. A lack of understanding of the impacts of dark tourism experiences upon tourists means that the dark tourist is thus often presented to society – within sensationalist media reports especially – as an individual who could be likened to Boorstin's 'superficial nitwit' tourist, who lacks the critical capacity to process dark histories in an in-depth manner.[16]

To reach an understanding of whether dark tourism has a 'valid' social role beyond providing entertainment or a day out for the tourist, insights into the dark tourism experience are essential. Crucially, understanding of their motives, as well as how they process the information to which they are subjected, is essential. Such in-depth insights into the experiences of dark tourists would be beneficial to both the academic field of study and the tourism industry. They would make it possible to draw more substantiated conclusions concerning whether dark tourism is an activity that has the capacity to increase morality.

Methods

An interpretive approach was adopted to conduct the research presented within this chapter. The empirical data presented here was drawn from a larger study of dark tourism experiences at four European sites. These sites included sites of warfare, genocide, grief and horror.[17] This chapter draws upon excerpts from in-depth interviews with four individuals who had recently visited the case study sites. The interviews were conducted using a combination of creative conversation and interactive interview techniques that promoted 'dialogue rather than interrogation'.[18] The research aimed to provide in-depth insights into the dark tourist experience. To give credence to participants' voices, each interview transcript was subjected to poetic-structure narrative analysis, which entails being attentive not only to what was said but also to how it was said, to discover factors that are most pertinent to research participants.[19] In using poetic-structure narrative analysis it is possible to observe how participants use linguistic devices such as metaphor, simile, tone and humour to accentuate certain points.[20] Within the next section, the narratives of four tourists are discussed to explore emergent research themes.

Dark tourist reflections on the role of dark tourism experience in everyday life

Common among the dark tourists within this study was that many sought insights into the past that might emerge through place-based learning. For example, Martin, a university lecturer who frequently visited dark tourism sites, travelled to Krakow, Poland, with the sole intention of visiting Auschwitz. He had previously visited many dark tourism sites, including sites associated with the Second World War as well as with the American Civil War. He relates this, at least in part, to his family history. His father had died in the Second World War, while he had been involved with the Territorial Army for many years and was also very interested in social history. Martin's status as something of a serial dark tourist is interesting given that it has previously been suggested that dark tourism is rarely the 'principal motive for travel, or a major activity at a destination'.[21] Martin explained that in the case of he and his wife, 'We are actually going to see Auschwitz, one feels almost embarrassed saying it, you know? There's almost some sort of voyeuristic, embarrassment, that's what you're going for, but that's what we are going for.' Martin's embarrassment concerning his desire to visit Auschwitz provides a glimpse into how he feels about his visit to a death site and his implicit gazing at the 'pornography of death'.[22] He seems self-conscious regarding how this decision will be perceived, in a society where death is either hidden or forbidden and where focusing upon death is seen as morbid,[23] a disrespectful intrusion upon grief or just plain weird.[24] Martin is aware of the charge of voyeurism brought against him as a dark tourist.[25] However, he makes 'no excuses' for his visit. In terms of the impact of the experience upon Martin's everyday life, he explains how he was apprehensive about how he might react to the visit beforehand, while the experience proved to be emotive for him. Following the visit, he left with a sense of sombre depression and gratitude, underpinned by a sense of relief and hope:

> Thank god it wasn't me. Thank god I wasn't in this, live through it, be part of it, or be, you know, alive during the Second World War. I think you have this hope that, it's over and it will never happen again and, you know, we've learnt from it . . . I mean you've got to feel like that, because the other side of it is just black depression (laughs) . . . I still, you know, you constantly think, how could that happen in Europe? How could that happen in Europe during the life of my parents?

Yet not all of those who visit dark tourism sites do so to learn from the past. For example, another participant, Elwyn, visited Auschwitz out of a sense of duty and reverence while visiting Krakow on holiday with his wife. It has been suggested that people often visit dark tourism sites because they are near to a dark site.[26] Yet, this explanation would undermine the deep reflexivity that Elwyn expressed both before, during and after the visit concerning his decision to visit. There is, therefore a need to look below the surface at the underlying reasons for Elwyn's visit. This is made even more apparent by the difficulty that Elwyn had in deciding whether to visit. He suggested that he first contemplated visiting, but then negotiated with his wife as to whether they should go or not. The final decision to visit came about because, as he put it:

> I thought I ought to go. I went because it was something that I felt one needed to see. I felt I was going there for the right reasons. It wasn't a pilgrimage, I've been on things called pilgrimages, what I would call pilgrimages, I've been to see, places where people have been buried, I've not been to places, other than the Somme, where people have been killed, and that was not the appeal. . . . It was a sense of duty. It was possibly a sense of reverence, I argued with myself a little bit: 'Should I be going there?' and I guess I'm not alone in that.

Elwyn stated that he was not drawn to the site because of the deaths that had occurred there and was keen to express that this was not the 'appeal' of the site. He, therefore, suggested that morbid curiosity did not act as a motivator:[27] he did not want to be at a site of tragedy, but felt he should be there. Ultimately, he believed that he was visiting to pay his respects, and indeed his behaviour on site came across as respectful and empathetic. Many authors have noted that remembrance and paying respect act as motivators for visitors to dark tourism sites. Tarlow even suggests that dark tourism may be a form of virtual nostalgia, in which case it could be suggested that Elwyn travelled to the Nazi era to heal the hurt of the past.[28] Silence and hushed tones dominated Elwyn's story; he noted the silence of people at the site and the reflective silence that he had enforced on himself following the visit, in only telling select tales to people he meets. He also saw it as significant that he wrote only one line in his photo album regarding the visit. At the end of the interview, he recalled:

> As we left to get back on the bus either I or Lyn said to the other: 'do you want to talk about it?' and whoever the other was said: 'no let's leave it for the time being'. We didn't say a word, I don't think on the way back for an hour, and there was virtually no conversation on that bus. It took the two of us, I think, a couple of days to sort it out in our minds. I still haven't sorted it out.

Such expressions of the desire and ability to empathise through dark tourism experiences were commonplace among all the research participants' narratives, even in what could be perceived as the most unexpected contexts. Sue was a 57-year-old American tourist who took part in a horror tour in London. In discussing why she chooses to participate in this and other dark tourism experiences, she stated:

> I think it's for me, it is kind of paying tribute to those who died . . . It's like, I've been to Dachau and Auschwitz and, and it's like letting those people live,

you know it's honouring them. I've been to many of the military cemeteries around Europe . . . and that is like this, you know, these people deserve to be remembered, the people who were killed. So it's for me it's not the sort of the bizarre whatever, but it's, you know, just think if it was one of your loved ones that was killed, you want them remembered.

Sue relates her empathetic feelings to her own situation. She is grateful for the sacrifices made for her. Therefore, she feels her motives should be respected and appreciated as legitimate. For Sue, the desire to remember past atrocities and the perceived value of doing so is what appears to motivate her to visit these sites.

Problematising the extent to which dark tourism sites offer opportunities for reflection

From the above discussion, it is clear that dark tourism experiences do have the capacity to nurture empathy, albeit perhaps among those who are predisposed to experience it. Yet it is also the case that experiencing does not always, in all contexts, inspire such depth of thought. For example, with the exclusion of notable sites, including Auschwitz or the Killing Fields in Cambodia, dark tourism – in most of its present guises – does not offer a tourist experience that starkly contrasts with an average day-trip, with the focus still placed upon the ability to 'step back in time' for a fun and thrilling experience. The similarities with an average tourist experience found at dark tourism locations can certainly appear distasteful. This is perhaps best demonstrated by tourists choosing to interact with sites through garish, commercialised aspects of the places, such as a 'Kodak moment' at the Bridge on the River Kwai, which allows tourists a perfect frame and viewing platform of the bridge while being sold the latest camera equipment. Visitors are also frequently observed exhibiting unreflective and arguably inappropriate behaviours; indeed, at many sites they are encouraged to do so. For example, a shooting range at the Cu Chi tunnels in Vietnam offers tourists the chance to fire a gun of their choice (AK-47s and Uzis are among those on offer). The site also claims to commemorate the memory of three million innocent Vietnamese civilians. Furthermore, Uzzell[29] has criticised approaches taken by dark tourism sites, such as Bovington Tank Museum and the Imperial War Museum, questioning the motives behind decisions to have warfare arcade games and items of replica military uniforms for sale in their museum shops.

There are many examples of dark tourism sites that limit critical reflection on everyday life. Indeed, the examples given above may be deliberate attempts to distance the tourist from the realities of tragic events occurring at the death sites visited. At such sites, the unreflective tourist is therefore enabled to return home feeling that they have experienced a piece of history. They are able to fix the place in time, while the horror of the tragedy that occurred in these places (in these instances, an estimated three million war dead in Vietnam and 100,000 dead in the building of a railway in Thailand) may well elude them. While the spaces themselves remain frozen, tied to the tourist industry that supports their continuing existence, they are lands with one certain future – one that will involve a continued reliving of the past – and in this respect they are dead places, in several dimensions.

A call to nurture compassion at dark tourism sites

Tourists clearly experience varying degrees of empathy following visits to dark tourism sites. Yet some of these individuals questioned whether this was enough in itself. For example, David, a vicar and former army chaplain from Birmingham, reflected on his experience of undertaking a First World War battlefield tour:

> On reflection, I found the whole thing sanitised, because of my army experience in the past. I thought this isn't, this is almost like voyeurism. This is watching other people, in this case suffer and die, although historically, but not actually being particularly discomforted. You know, we went back home to that wonderful hotel, and I had snails and ice-cream at one o'clock in the morning and then got up the next morning, and then sat on another battlefield where thousands of people had died.

On reflection almost a year after visiting the battlefields, David had been deeply affected by the tour emotionally, which had spurred a further quest to explore the First World War via an on-foot battlefield tour. In the above narrative he identifies a key debate concerning whether dark tourism has similarities to pilgrimage and is therefore an honourable, respectful activity,[30] or whether such interests should be perceived as voyeuristic.[31] David recognises himself as a battlefield tourist and notes that remembrance is central to his motivation, and yet he is also unsure of the purpose that dark tourism serves in the present context.

The sense of relief felt by many tourists leaving death sites – the 'thank god it's not me' effect that Martin spoke of – is perhaps not enough to ensure that dark tourism contributes towards peace through tourism.[32] Tourists are able to get back on their tour bus safe in the knowledge that they are returning to the more comfortable, sequestrated space and time of their present. Dark tourism sites, particularly the various dungeons and chambers of horrors that depict medieval history, often display an awful past and reflect on present-day improved situations. Opportunities to reflect on the past and its relationship to the present day are rare. As a result, tourists leave these spaces satisfied with a representation of the singular past; rarely, however, will they leave with an appreciation of how the trajectory of history connects to their everyday, less still to a better future for humanity. As the accounts above reveal, they are more likely to feel a sense of gratitude for not having had to endure the suffering of those who suffered at these sites, while among the tourist accounts there are no expressions of a desire to create change within the world, to help those who are suffering in similar conflicts and tragedies. As David notes, visiting death sites cannot change the past; therefore, dark tourism might be perceived as a futile activity.

What changes might ensure that a dark tourism site has greater pro-social benefits? It is argued here that dark tourism sites could be in a far stronger position to ensure the aim of 'never again' if they were to nurture compassion, rather than encouraging the emotional resonance that arises from empathy alone. Compassion is a development upon an initial feeling of empathy, arising because of a desire to do something with the emotions experienced at dark tourism sites. Examples of actions based upon empathy might include, for instance, involvement in campaigning, charity or other efforts that centre upon the removal of human suffering or raising awareness of the risks of prejudice (in the case of human atrocities against fellow humans). Individuals who are affected by visiting sites of natural disasters might think about their role in tackling environmental crises following their visits. After visiting a warfare site, they might consider more deeply the importance of non-violent forms of conflict resolution. Encouraging individuals to feel compassionate as a result of visiting dark tourism sites would enable such sites to claim a tangible contribution to peace via the examples given above.

Many of the participants within this study expressed, following their visit, that they had had difficulty in knowing what to do with the knowledge and emotional experiences they had gained. Further, while several mentioned that they were going to visit dark tourism sites again, none stated that they were going to participate in what might be described as compassionate action because of their experiences. Yet, encouraging empathetic responses would potentially also be emotionally rewarding for the tourist. As the Dalai Lama states:

> being compassionate does not mean remaining entirely at the level of feeling, which could be quite draining ... Compassion means wanting to do something to relieve the hardship of others, and this desire to help, far from dragging us further into suffering ourselves, actually gives us energy and a sense of purpose and direction.[33]

A means by which to encourage empathy through dark tourism would be on-site interpretation. Uzzell made the case for such efforts at the beginning of the 1990s,[34] stating that dark tourism sites were currently not taking full advantage of opportunities to expose the links between the past, present and future. Doing so, he argued, would inform visitors about the historical processes that had affected their lives; such interpretation would 'empower them, by enhancing their capacity to understanding, to perhaps change, their world'. For Uzzell, effective interpretation of a traumatic past involves 'hot interpretation', which implies going beyond the reflective spaces provided at many sites. Hot interpretation gives individuals valuable opportunities to consider what they have witnessed. It involves exposing previous generations' grave moral errors, as well as perhaps exposing the potential for repetition in contemporary society. Such interpretation would encourage at least a state of vigilance, if not active participation among modern-day visitors. Further still, Uzzell suggests that hot interpretation may be a means of leading visitors towards 'hot action'.

An example of such hot interpretation might be observed at Anne Frank Haus, Amsterdam. Conversely to most contemporary interpretive techniques provided by the majority of dark tourism sites, some of the displays at Anne Frank Haus provide some signposts for journeying towards a form of dark tourism that attempts to nurture an empathic response through hot interpretation. Following a visit to the house, the tourist enters a visitor centre where they are invited to take part in an interactive game that allows them to make connections between expressions of contemporary prejudices and their potential to lead to major catastrophe. The exhibition encourages the visitor to confront moral choices made in the present day by connecting to current issues, for example, football hooliganism and homophobic song lyrics. The tourist watches a series of video clips from arenas including popular culture and sporting events. The visitor is then asked to respond quickly to a difficult question about these representations. Following this, they are given the opportunity to reflect upon their answers. This display allows the tourist to connect instantly with the resonance of the piece of history portrayed through the house. It allows an appreciation of how such battles are not 'in the past' and therefore held at a safe distance. Rather, the tourist learns that to maintain human rights, there is a need for constant personal and collective reflection. This exhibit also enables the tourist to appreciate the perspective of the bystander, one that is commonly omitted from the telling of dark histories. In so doing, the tourist is perhaps able to grasp a sense of endless humanitarian responsibility in their present-day life.

A question remains: would the tourist actually want this form of experience? Locating dark tourism spaces in the context of contemporary realities might be deemed too difficult to deal with for those trying to enjoy their holiday or for the person looking for an escape route from the present. Indeed, with compassion comes a sense of responsibility. The avoidance of that responsibility is perhaps part of the appeal of the tourist experience. The implications of such thinking for the industry include its need to recognise that it has a responsibility not just to its immediate local community, to survivors and to the relatives of victims, but – in an increasingly globalised world – to global citizens, who may or may not choose to visit such sites. Massey refers to the perspective alluded to here as a 'politics of outwardlookingness, from place beyond place'.[35] In discussion of this concept, she states:

> Neither space nor time can provide a haven from the world. If time presents us with the opportunities of change and (as some would see it) the terror of death, then space presents us with the social in the widest sense: the challenge of our constitutive interrelatedness; the radical contemporaneity of an ongoing multiplicity of others, human and non-human; and the ongoing and ever-specific project of the practices through which that sociability is to be configured.[36]

From a tourist's perspective, the development of a compassionate viewpoint would involve doing a little more reflection before getting back on the bus, to see the dark events as part of a global, interlinked world, as a first step. It would involve a harnessing of both the relational thinking and the outwardlookingness that Massey describes to identify means of operating within the contemporary world to enact change.

Conclusions

Within this chapter the phenomenon of dark tourism has been confronted as meaning many things to many people – a day out, a memorial, a source of income for communities, to name but a few. Here we return to the original question, which concerns whether dark tourism sites enable critical reflection upon everyday life, the assumption being that doing so might result in creating a cultural shift to enable dark sites to validly claim a contribution to 'peace through tourism'. In exploring the in-depth experiences of dark tourists it is possible to observe that dark tourism experiences do lead to expressions of empathy, as well as to increases in understanding – to the desire to know more, and a desire to benefit personally (emotionally, cognitively, spiritually) from such experiences. Yet, we have also seen that the focus upon memorialisation within dark tourism creates a certain response from tourists. Site narratives typically leave tourists at a point of liberation or resolution of the past. At the end of a visit to Auschwitz-Birkenau, for example, one is left to confront a memorial stone for the murdered. The final display at the S-21 prison in Cambodia is a memorial to a particularly heinous murder of a group of women prisoners, while one exits the Cu Chi tunnels in Vietnam via the gift shop, where one can buy scarves or bullets as souvenirs of the past. Arguably, as a result, dark tourists leave with a sense that the events depicted within sites are now safely 'in the past'. They could therefore be forgiven for thinking that there was little need for action on their part within the present day.

Yet, as has been noted, atrocious events continue to occur within the modern world, despite the existence of dark tourism sites, history lessons and popular culture depictions of past atrocities. It is clear that halting human suffering through dark tourism is likely to be an impossible ideal. Yet it may be possible for such sites to make a far greater contribution towards peace through tourism if they attempt to go beyond empathy and nurture compassion through employing interpretive strategies. A current notable example of such a strategy would be that of the exhibitions that form part of Anne Frank Haus, which seek to contribute to human rights education. It has been argued in this chapter, therefore, that there is potential for dark tourism sites to nurture compassion through their presentation of dark events in history. This, it has been argued, would extend the ethics of care of sites to include diverse groups of survivors, visitors and local and global citizens by seeking to create representations of history that are more reflective of ongoing histories, power dynamics of place and multiple trajectories of those spaces. This would extend the significance of dark tourism by focusing upon empowerment for those who visit, as well as a relationship with the communities in which these places reside – in both space and time.

In the face of significant twenty-first-century challenges, including global warfare and environmental crises, this is a compelling proposition for the future development of such sites. As many key sites become more chronologically distant from us – for example, the First World War has now passed from living memory – dark tourism sites could do far more than provide memorials. They arguably now have a role in ensuring in a more active sense that the lessons of the past influence societal progression in the future. If this were to happen, it is possible that we would witness the emergence of an industry that could help to shape societies in a more meaningful way – for if there is to be a genuine social benefit from dark tourism, it is to be found in the dynamic contemporaneity of these spaces: in how people pass through them, and are influenced to instigate change because of them, in their present day. Occupying such space would perhaps be a little more uncomfortable for tourists, and indeed those tasked with interpretation. It might lead to argument, contestation and confusion, but this may also enable societies to live a little differently. This would involve engaging with dark tourism spaces as geographies of ongoing identity construction, where interconnectedness and the social are central.

Notes

1. A. Pritchard, N. Morgan, and I. Ateljevic, "Hopeful Tourism: A New Transformative Perspective," *Annals of Tourism Research* 38, no. 3 (2011): 941–63.
2. M. Foley and J. J. Lennon, "JFK and Dark Tourism: A Fascination with Assassination," *International Journal of Heritage Studies* 2, no. 4 (1996): 198–211.
3. C. Strange and M. Kempa, "Shades of Dark Tourism: Alcatraz and Robben Island," *Annals of Tourism Research* 30, no. 2 (2003): 386–405.
4. P. Levi, "Revisiting the Camps," in *The Art of Memory: Holocaust Memorials in History*, ed. J. E. Young (New York: Jewish Museum with Prestel-Verlag, 1986), 185.
5. A. Schwabe, "Visiting Auschwitz, the Factory of Death," January 27, 2005, accessed June 5, 2007, www.spiegel.de/international/0,1518,338815,00.html.
6. P. Corbetta, *Social Research: Theory, Methods, Techniques* (Thousand Oaks, CA: Sage, 2003); R. K. Schutt, *Investigating the Social World: The Process and Practice of Research* (Thousand Oaks, CA: Sage, 2006).
7. P. Williams, *Memorial Museums: The Global Rush to Commemorate Atrocities* (Oxford: Berg, 2007).
8. J. Schofield, "Monuments and the Memories of War: Motivations for Preserving Military Sites in England," in *Material Culture: The Archaeology of Twentieth Century Conflict*, ed. J. Schofield, W. G. Johnson, and C. M. Beck (Routledge: London, 2002), 5–24.
9. E. Lee-Potter, "Thousands of Teenagers to Visit Battlefields of the First World War in New Government Scheme," *The Independent*, October 22, 2014.
10. Centenarybattlefieldtours.org, accessed November 3, 2015.
11. V. L. Smith, "War and Tourism: An American Ethnography," *Annals of Tourism Research* 25, no. 1 (1998): 202–27.
12. D. Uzzell, 'The Hot Interpretation of War and Conflict', in *Heritage Interpretation, Vol. I. The Natural and Built Environment*, ed. D. Uzell (London: Belhaven Press, 1989), 33–47.
13. J. J. Lennon and M. Foley, *Dark Tourism* (London: Cassell, 2000).
14. K. Rupesinghe, *Civil War, Civil Peace* (London, Pluto Press, 1998).
15. P. Williams, *Memorial Museums: The Global Rush to Commemorate Atrocities* (Oxford: Berg, 2007).
16. D. J. Boorstin, *The Image: A Guide to Pseudo-events in America* (New York: Vintage, 1987).
17. R. A. Dunkley, N. Morgan and S. Westwood, "A Shot in the Dark? Developing a New Conceptual Framework for Thanatourism," *Asian Journal of Tourism and Hospitality Research* 1, no. 1 (2007): 54–63.
18. C. Ellis and L. Berger, "Their Story/My Story/Our Story: Including the Researcher's Experience in Interview Research," in *Handbook of Interview Research*, ed. J. F. Holstein and J. A. Gubrium (Thousand Oaks, CA: Sage, 2002), 851.

19 C. K. Riessman, *Narrative Analysis* (London: Sage, 1993).
20 J. P. Gee, "A Linguistic Approach to Narrative," *Journal of Narrative and Life History* 1, no. 1 (1991): 15e39.
21 A. V. Seaton and J. J. Lennon, "Thanatourism in the Early 21st Century: Moral Panics, Ulterior Motives and Alterior Desires," in *New Horizons in Tourism: Strange Experiences and Stranger Practices*, ed. T. V. Singh (Wallingford: CABI, 2004), 69.
22 G. Gorer, "The Pornography of Death," in *Death, Grief, and Mourning*, ed. G. Gorer (London: Cresset Press, [1955] 1987).
23 P. Aries, "The Hour of Death," in *Death, Mourning and Burial: A Cross-cultural Reading*, ed. A. C. Robben (Oxford: Blackwell, 2004), 40–9; E. Kubler-Ross, *On Death and Dying* (New York: Macmillan, 1976).
24 G. J. Ashworth, "Tourism and the Heritage of Atrocity: Managing the Heritage of South African Apartheid for Entertainment," in *New Horizons in Tourism: Strange Experiences and Stranger Practices*, ed. T. V. Singh (Cambridge: CABI, 2004), 95–108.
25 T. Blom, "Morbid Tourism – A Postmodern Market with an Example from Althorp," *Norsk geogr. Tidsskr* 54 (2000): 29–36.
26 Lennon and Foley, *Dark Tourism*.
27 G. M. S. Dann, "Tourism: The Nostalgia Industry of the Future," in *Global Tourism*, ed. W. F. Theobald (Oxford: Butterworth-Heinemann, 1995), 35–7.
28 P. Tarlow, "The Appealing Dark Side of Tourism and More," in *Niche Tourism: Contemporary Issues, Trends and Cases*, ed. M. Novelli (London: Elsevier, 2005), 47–57.
29 Uzzell, "The Hot Interpretation of War and Conflict," *Heritage Interpretation* 1, no. 35 (1989): 35.
30 T. Walter, "War Grave Pilgrimage," in *Pilgrimage in Popular Culture*, ed. I. Reader and T. Walter (Basingstoke: Macmillan, 1993): 63–91.
31 G. M. S. Dann, "There's No Business like Old Business: Tourism: The Nostalgia Industry of the Future," in Theobald, ed., *Global Tourism*, 35–7.
32 Ibid.
33 XIV Bstan-'dzin-rgya, D. L., *Beyond Religion: Ethics for a Whole World* (Croydon: Random House, 2012), 55.
34 Uzzell, "Hot Interpretation of War and Conflict," 36.
35 D. Massey, *For Space* (London: Sage, 2005), 192.
36 Ibid., 194.

9 The undead and dark tourism
Dracula tourism in Romania

Duncan Light

Introduction

Although I have been researching Dracula tourism (the visiting of places in Transylvania associated with the Count Dracula of fiction and cinema) for more than a decade, I have not previously examined it in terms of dark tourism since, to my mind, there was little direct connection between such tourism and death or human suffering. However, conceptions of dark tourism are continually evolving and the phenomenon is now increasingly defined in terms of the 'macabre', meaning that dark tourism now embraces a wider range of sites and experiences which are not directly associated with death. I wish to begin this chapter by considering the changing definitions of dark tourism, with particular reference to the notion of the macabre as a criterion for defining the phenomenon. I then want to examine Dracula tourism in Romania from the perspectives of both supply and demand. In terms of supply I argue that there is almost no deliberate provision of tourist experiences based on Dracula, so that identifying Dracula tourism as a form of dark tourism is problematic. When looking at demand, however, I argue that Dracula tourism is a heterogeneous phenomenon which encompasses a wide range of interests and expectations among tourists.

The changing scope of dark tourism

Early definitions of dark tourism defined the phenomenon as 'the presentation and consumption (by visitors) of real and commodified death and disaster sites'.[1] Similarly, the related concept of thanatourism was defined as 'travel to a location wholly, or partially, motivated by the desire for actual or symbolic encounters with death, particularly, but not exclusively, violent death'.[2] As many others have pointed out, these two concepts had different emphases: dark tourism was defined by a supply-led perspective, while thanatourism adopted a demand-led approach.[3] However, both definitions placed death – in particular, human death – squarely at the centre of this form of tourism. More recent conceptions of dark tourism (although not thanatourism), though, are rather broader. One influential definition states that dark tourism is 'the act of travel to sites associated with death, suffering and the seemingly macabre'.[4] This raises the question of how dark tourism (hitherto focused quite specifically on death) has now come to embrace the 'seemingly macabre'? This change in scope seems to have been a response to various 'strange' or unusual forms of tourism (most notably, Gunther von Hagens' 'Body Worlds' exhibition) that are less directly connected with death, and which do not fit easily into existing typologies or definitions of dark tourism.[5] Presumably it is the associations with the macabre that have led recently to Dracula tourism being considered as a form of dark tourism. However, since Dracula tourism does not fit easily into Seaton's typology of thanatourism sites, it has not been identified as a form of thanatourism.[6]

The Oxford English Dictionary defines 'macabre' as 'grim, horrific or repulsive'.[7] Clearly, many places/sites with such properties are directly associated with death. However, this is not always the case: some places that are labelled as dark tourism sites are not especially grim or horrific and have few direct associations with death. Examples might include parks of communist-era statues in central/eastern Europe, active volcanoes, Cold War nuclear bunkers, decommissioned nuclear power stations, the birthplaces of communist leaders and sites of weapons testing.[8] The macabre is especially problematic because it is a normative judgement with implicit overtones of aberrance, deviance and transgression: that which is macabre somehow flouts social norms. This links to Seaton's argument that the very term 'dark tourism' is a judgement based on an implicit contrast with 'light tourism'.[9] In addition, the macabre is a social construction, so that its meaning is culturally and individually relative, rather than absolute.[10] The macabre will of course mean different things to different people, and that which societies deem to be macabre varies considerably over time and space. As such, labelling something as macabre (and, consequently, dark) is 'a complicated matter of perspective and privilege'.[11]

As the scope of dark tourism continues to broaden, almost anything that is somehow associated with death (or the macabre) is now being considered as dark tourism.[12] Indeed, some recent conceptualisations of the topic have argued that dark tourism need not include death at all![13] Consequently, Sharpley has argued that 'such is the variety of sites, attractions and experiences now falling under the collective umbrella of dark tourism that the meaning of the term has become increasingly diluted and fuzzy'.[14] This leaves dark tourism open to the criticism that the term fails to differentiate between very different types of site, based on very different associations with death, and offering completely different experiences to visitors.[15] In response, various models have attempted to clarify the highly variegated nature of dark tourism. Stone proposes a spectrum of dark tourism supply, ranging from the darkest sites (sites of death or suffering with a predominantly educational orientation) to the lightest (those which are associated with, or represent, death or suffering and which are orientated predominantly around entertainment).[16] Dracula tourism is implicitly identified as one example of 'lightest dark tourism'.[17] While this model is useful for differentiating dark tourism provision (and the intentions behind such provision), it does not consider the experiences of visitors to dark attractions and the extent to which an encounter with death shapes their reasons for visiting.

Sharpley has also sought to clarify the nature of dark tourism by proposing a model which identifies four 'shades' of dark tourism.[18] These are based on the nature of supply (accidental or purposeful) and demand (the extent to which visitors are seeking an encounter with death). This is important in recognising that both the supply of and demand for dark tourism are extremely heterogeneous, and it also acknowledges that tourists may have a wide range of reasons for visiting dark tourism places. However, it is widely recognised that the whole issue of why people visit dark tourism attractions is currently poorly understood. Moreover, there is a growing acceptance that not everyone who visits such places does so due to an interest in death (or the macabre),[19] meaning that many of the people at so-called dark attractions may not, in fact, be dark tourists.[20] To better understand dark tourism there is, therefore, a need to look in detail at the motivations, expectations, experiences and practices of the tourists who participate in it. In particular, there is a need to engage with how such motives and experiences differ between different types of dark tourism attraction, but also within an individual attraction.[21]

The dark in Dracula tourism

In terms of supply, there is little attempt in Romania to provide attractions or experiences for Dracula tourists. Romania has long been reluctant to cater for Dracula tourism, since the associations with vampires and the supernatural fundamentally collide with Romania's sense of its cultural and political identity.[22] Consequently, for more than forty years, Dracula has barely featured in Romania's state-sponsored tourism promotion. Instead, the country has emphasised other forms of tourism in Transylvania, such as cultural and heritage tourism, mountain tourism and, more recently, ecotourism/ wildlife tourism, outdoor activities and culinary tourism. The state's approach to Dracula tourism has been to tolerate rather than encourage this form of demand.[23]

Consequently, the places in Transylvania that are mentioned in Bram Stoker's novel can all be visited by Dracula enthusiasts, but most have made no attempt to cater for their interest. For almost five decades Dracula tourists have sought Castle Dracula, but since the castle never existed outside Stoker's imagination their searches have been in vain. This proved not to be a deterrent, since another Transylvanian castle – Bran Castle – was quickly appropriated as Castle Dracula (see Figure 9.1). Bran has no connection with the fictional Castle Dracula (and is situated 150 miles south of where Stoker placed the castle) and does not feature in any of the Dracula films; however, as an imposing collection of spires and turrets, situated on a hilltop and located in Transylvania, Bran Castle *looked* right. Consequently it was quickly labelled by Western tourists as 'Dracula's Castle' and became, and remains, one of the major sites of Dracula tourism in Romania. That said, the castle's managers have never done anything to encourage the association with Dracula. Indeed, since opening as a visitor attraction in 1957 the building has been presented as a museum of medieval art, and more recently as a royal holiday home, although outside the castle a thriving market in horror-related souvenirs has now developed, and a private entrepreneur has opened a 'Castle of Horror' (a small indoor attraction which offers a short experience of a recreated haunted castle).

Figure 9.1 Bran Castle in southern Transylvania (frequently mistaken for Castle Dracula).
Photo copyright D. Light.

Elsewhere in Transylvania, a hotel (now called Hotel Castle Dracula) was constructed in the 1980s in the approximate part of the Carpathians where Bram Stoker placed his fictional Castle Dracula. The hotel vaguely resembles a castle and includes a tower, crenelated walls and an interior courtyard, while the reception and dining areas feature dark wood panelling and stuffed animal heads in an attempt to recreate the ambiance of spooky Gothicism. The hotel's managers have also established a crypt in an underground room which tourists can visit for a small fee, and which features vaguely macabre wall paintings (inspired by Dracula films) and a coffin in the middle of the room. A bored hotel employee recites the story of Dracula and his brides and then unexpectedly turns out the lights, whereupon another figure leaps from the coffin, generally producing shrieks from visitors. This is certainly an attempt by the hotel to produce a commodified experience, based on horror and the macabre – although there is not much else in the hotel that is macabre, and I am not convinced that an (optional) experience which lasts for less than five minutes makes the hotel a site of dark tourism. In short, there is little deliberate provision of Dracula-related experiences in Transylvania so that, in terms of supply, identifying Dracula tourism as a form of dark tourism is problematic. Only Bran's 'Castle of Horror' and Hotel Castle Dracula's short crypt experience are commodified representations of the macabre that could explicitly be identified as forms of 'lighter' dark tourism supply. For the most part, Dracula tourists have to make do with the dramatic landscapes of Transylvania, or (as I argue later) make their own experiences of Dracula by using their imaginations.

In terms of demand, Dracula tourists are not a homogeneous group. Instead, Dracula tourism embraces a range of interests and motives, each associated with different degrees of interest in the dark and macabre (and sometimes no interest at all). One of the longest established motives for visiting Romania is a search for the literary roots of Bram Stoker's *Dracula*. The novel created a powerful place myth of Transylvania as 'one of the wildest and least known portions of Europe'.[24] Stoker's Transylvania is a remote, sinister and frightful place, untouched by Western science and rationality, where vampires roam freely. Stoker portrayed Transylvania as recognisably European (Count Dracula is, after all, a Hungarian aristocrat who speaks fluent German), but at the same time sufficiently different (and horrifying) to be threatening to the 'civilised' West. Transylvania has long attracted fans of Gothic fiction – and *Dracula* in particular – wanting to see for themselves the landscapes and settings of the novel, an activity which can be understood as a form of literary pilgrimage.[25] For example, the British 'Dracula Society' has visited Transylvania regularly since 1974, while many other fans have made individual visits or have joined organised tours to the region. Several of the tourists whom I interviewed on one Halloween tour to Transylvania explicitly spoke of wanting to follow in the footsteps of Jonathan Harker (an Englishman who journeys to Castle Dracula in the early part of Stoker's novel). However, literary tourists are themselves a diverse group and are not confined to fans of *Dracula*. The broader genre of vampire fiction has enjoyed extraordinary popularity in recent decades and although many of these novels do not mention Transylvania, the region has an established reputation as the seemingly natural home of vampires. This, in turn, draws enthusiasts wanting to see Transylvania for themselves. Another type of literary tourist is those seeking inspiration for their own writing; on the Halloween tour that I joined, two of the twenty-one participants were intending to write their own vampire novels and were visiting Transylvania to gain local knowledge and colour.

Is there anything about such literature-based Dracula tourism that constitutes dark tourism? After all, the rather earnest motives assumed of literary tourists seem far removed from the 'lighter dark tourism' of which Dracula tourism is apparently an example. There is certainly plenty about death in Bram Stoker's *Dracula*: Count Dracula has many victims among the Transylvanian (and British) population and meets his own death at Castle Dracula at the end of the novel.

More broadly, a visit to Transylvania is underpinned by the absent presence of Count Dracula, himself a monstrous (and macabre) figure. In this sense literary tourism in Transylvania could be loosely considered as tourism based on death and the macabre. This, however, raises the question of whether dark tourism can embrace death that has taken place in fiction. Some authors find this unproblematic, and in his discussion of 'dark fun factories' Stone argues that fictional death can be the basis for dark tourism experiences.[26] However, I am not convinced: in my view, to include death that is represented in fiction under the umbrella of dark tourism rather stretches the concept beyond its limits. In any case, I suspect that many Dracula tourists seeking the literary origins of *Dracula* are more interested in the dramatic landscapes of the region and in particular the setting of Castle Dracula, rather than Dracula's (fictional) death.

A second, related, form of Dracula tourism is the desire to see a place – Transylvania – that has previously been encountered through film. This can be considered as a form of film-induced tourism (although since more people have encountered Dracula films on television, the term 'screen tourism' is perhaps more appropriate).[27] More than 350 films have been made about Count Dracula and most feature Transylvania in some way. Screen tourists (like literary tourists) are drawn to the place myth of Transylvania, but especially to the sinister and terrifying Castle Dracula situated high in the Carpathian Mountains. A well-established characteristic of screen tourism is that tourists seek the location that is represented in a film rather than the location where filming actually took place.[28] None of the major Dracula films were made in Transylvania (and Hammer Studios famously used woodlands near its Berkshire base to represent the forests of Transylvania). Nevertheless, screen tourists are eager to see for themselves the places in Transylvania portrayed in the Dracula films.

Both literary and screen Dracula tourists are seeking an encounter (albeit a light-hearted one) with the horror of the Dracula story. In this, there are clear parallels between Dracula tourism and what has been termed 'Gothic tourism'.[29] This involves tourism experiences which are about engaging with (and are underpinned by) the tropes and discourses of Gothic fiction. It is a ludic form of tourism with a strong affective dimension based on horror, dread, thrills, frights and a delight in the uncanny and unknown. In particular, Gothic tourism involves a theatrical delight in creating or entering imaginary (or imaginative) worlds. Certainly, for many screen (and literary) tourists, a visit to Transylvania includes a strong imaginative and playful component. Being in a location portrayed in a film is an opportunity to suspend disbelief and 'connect' with the imaginative world of that film through engaging in fantasy, escape and dreamwork. For example, Dracula tourists may hope to encounter Count Dracula (in some form) during their visit to Transylvania. At Hotel Castle Dracula, some enthusiasts whom I interviewed went outside on solitary searches for bats or wolves.[30] When visiting film locations, tourists often engage in imaginative play by re-enacting iconic scenes from the film itself.[31] This is quite common among Dracula tourists: at Bran Castle, visitors often play at being Dracula, adopting the classic stance of arms aloft ready to pounce on a victim and intoning lines from the film.[32] Others hide behind doors or furniture and leap out on their family or friends with a blood-curdling shriek. In a variety of ways, tourists enchant their visit through blurring the boundary between the real and the imaginary.

Does any of this represent a form of dark tourism? A visit to a place associated with a horror film may be motivated by a desire to connect with something 'dark' (such as the castle where Count Dracula met his death). And without question, some screen (and literary) tourists are seeking an experience based on terror and dread when they visit either Bran Castle or Hotel Castle Dracula. However, this is quite different from entertainment-based dark tourism experiences where, for example, horror or death is represented in front of visitors. Instead, the horror in Dracula tourism is something that visitors have to make for themselves, based on what they bring with them from home, and by giving free rein to their imaginations.[33] Moreover, while such experiences may appear to be based on horror, those who take part do not appear to find them horrific (in contrast to entertainment-based representations of death and suffering). This illustrates the difficulties of using the 'macabre' as a criterion for defining Dracula tourism. The macabre means very different things to different people in different circumstances: those activities which observers may judge as rooted in the macabre may not be considered in such terms by the people taking part.

A further group of Dracula tourists (some of whom are also literary or screen tourists) are those drawn to the supernatural roots of the Dracula story. Transylvania is inextricably associated with vampires in Western popular culture (even though vampires are unknown in Romanian folklore), which is why some tourists are seeking an encounter with the supernatural during their visit. Transylvania is not an isolated case: the supernatural (and ghosts in particular) has long fascinated tourists.[34] Indeed, Bristow and Newman propose something called 'fright tourism' as a specific form of dark tourism/thanatourism.[35] Such fright tourism usually involves encounters with ghosts and the supernatural and offers experiences that are scary and thrilling, but which (unlike risk recreation or adventure tourism) do not involve actual danger. Related phenomena include 'ghost tourism',[36] 'spook tourism'[37] and 'haunting tourism'.[38]

Certainly there are a number of locations around the world that have a healthy tourism industry based on their associations with the supernatural and otherworldly. For example, Scotland has long attracted visitors in search of ghosts, spooks and phantoms; indeed, ghosts feature prominently in the country's tourism promotion.[39] Numerous cities throughout the world now offer ghost walks.[40] In particular, New Orleans has long appealed to tourists on account of its associations with the supernatural and occult.[41] The city offers visitors a range of guided walking tours themed around ghosts, cemeteries, Voodoo and vampires.[42] Similarly, Salem in Massachusetts now attracts over a million visitors a year due to the Salem witch trials of 1692, after which twenty people were executed for witchcraft. There are also places outside Romania that draw tourists because of their Dracula associations. The best example is Whitby in Yorkshire, which offers a 'Dracula Experience' attraction and Dracula-themed walking tours. Since 1994 the town has also hosted Gothic festivals (in April and at Halloween), and the associations with Dracula and the spooky are increasingly prominent in the town's tourism marketing.[43]

Around half of the group on the Dracula tour I accompanied stated that they had some sort of interest in the supernatural. But the nature of this interest varied considerably, and most simply enjoyed the supernatural as entertainment: their interest was in the pop culture vampire rather than in the occult. For example, one Australian tourist argued: 'I think everyone on this tour ... has come onto it because they love vampire fiction, not because they believe vampires exist'. An American tourist (who was a committed Christian) stressed that her interest in the supernatural was 'purely entertainment, you know what I mean; it's a little fantasy world and it's entertainment and, you know, it's fun'. Another visitor from America explained: 'I'm into horror and spooky stuff, I love being scared', but emphasised that it was all about fun. A Canadian visitor put it more bluntly: 'I would never come on this trip if I actually believed [in the supernatural].'

Central to the enjoyment of the supernatural as entertainment is the affective thrill of being 'spooked'. This involves a playful willingness to give freedom to the imagination and delight in the possibilities of the unknown and ineffable. It is an experience based on 'a reflective self-awareness of delusion and make-believe . . . a willing and deliberate suspension of disbelief'.[44] Another important aspect of being 'spooked', however, is a sense of fear and apprehension of that which is not known: vampires, ghosts and other supernatural creatures. At the same time this fear is mediated by the knowledge that the danger is not real and poses no actual threat. This was expressed by a Canadian visitor who said: 'I'm an atheist ... so I really don't believe in anything that goes bump in the night... I don't believe in it [but] I enjoy it ... fear'. This illustrates how the encounter with the imaginary-supernatural can offer intense but controlled (and temporary) thrills, based on 'pretend' danger.[45] Such 'delight without delusion'[46] involves only a suspension of normality, and the tourist is confident that they will be able to return safely and without difficulty to the 'real' world.[47]

At first sight, it might appear straightforward to label anything involving the supernatural as macabre, and therefore an example of dark tourism. But again, those Dracula tourists who engaged in such ludic and light-hearted enjoyment of the 'pop culture supernatural' may not consider it to be in any way macabre. The experience is again grounded in imaginative play and the enjoyment of what is unknown (searching for the monster when visiting Loch Ness might be another example). The success of the Harry Potter books and films testifies to the popularity of the supernatural as entertainment. And, with its emphasis on thrill, controlled danger and an ultimate return to safety, it could be argued that enjoyment of the pop culture supernatural has more in common with a rollercoaster ride than with dark tourism. Indeed, many theme parks feature rides (such as ghost trains or haunted houses) with distinctly Gothic themes, and their users are probably concentrating more on fun and entertainment than on death or the macabre.[48]

However, other Dracula tourists have a deeper interest (and sometimes a sincere belief) in vampires and the supernatural. Once again, the nature of this interest varies considerably. On the tour I accompanied, one person was deeply curious about vampires and their reputation for immortality; another enjoyed reading about the supernatural but did not want to go further. Several others spoke of how the supernatural was important and meaningful to them. For example, one couple stated 'we're in touch with our dark side', while another claimed to have 'been an occultist ever since I was 12'. While these visitors were happy to enjoy the supernatural as entertainment along with the other visitors, what had brought them to Transylvania seemed to be a search for some sort of connection – and contact – with the supernatural. This was particularly the case at Halloween, when the boundaries between the material and the spiritual worlds are believed to temporarily dissolve so that ghosts and other supernatural beings can reign more freely.

Such an interest in the supernatural might appear to be an unproblematic form of dark tourism on account of its associations with death and the macabre. Certainly some scholars have no difficulty in accepting travel to places associated with 'other', New Age or pagan spiritualities under the umbrella of dark tourism.[49] However, I again argue that the situation is not so straightforward. First, the word 'supernatural' is extremely wide-ranging, embracing anything and everything that transcends nature or scientific understanding.[50] Even a more narrow definition of the supernatural which focuses on the occult or paranormal embraces a wide range of beliefs and practices. In terms of dark tourism, not everything associated with the supernatural has connections with death, and certainly some supernatural entities are very much alive. There are parallels with Sharpley's claim that many of the people who visit the graves of famous people are more interested in their lives than in their deaths.[51]

Furthermore, defining touristic interest in the supernatural as macabre is also problematic. As I argued earlier, the macabre is a normative judgement made from a particular standpoint. Certainly those tourists with an interest in the supernatural and occult would not necessarily consider these topics to be macabre. In fact, for such tourists, a visit to Transylvania probably has more in common with religious/ spiritual tourism than dark tourism. It is a journey based on spirituality: something which 'assumes the existence of the supernatural, though not necessarily a god or gods, and therefore represents a wider connotation of the sacred'.[52] In particular, for some, the trip to Transylvania can be considered as a form of pilgrimage: travel redolent with meaning, to a place considered sacred within an individual's belief system, that is underpinned by a desire to connect with the Other.[53] It is a journey which provides an opportunity for connection and transformation, but which can also be an occasion to discover or express a new (or 'real') sense of identity.[54]

Conclusion

In this chapter I have examined the relationship between Dracula tourism and dark tourism and have considered if it is appropriate to identify Dracula tourism as a form of dark tourism based on existing typologies and ways of classifying the subject. In terms of supply, there is little basis for describing Dracula tourism (at least in Transylvania) as dark tourism. Romania has long been unwilling to cater for Dracula tourists and there is little provision in Transylvania of the commodified and entertainment-based experiences which constitute 'lighter dark tourism'. Bran Castle may be consumed as Castle Dracula by visitors, but the site's managers have never done anything to encourage the association. Hotel Castle Dracula *is* intended to cater for Dracula enthusiasts but, with the exception of the short 'crypt experience', there is little about the site that is dark or macabre. However, Dracula tourism takes different forms in different places and at other locations on the Dracula trail – particularly Whitby – there is more of a deliberate attempt to provide experiences for visitors based on death, horror and the macabre.

When looking at demand for Dracula tourism, the situation is much more complex. I have identified three broad (but overlapping) motives for taking part in Dracula tourism: an interest in the literary, cinematic and supernatural dimensions of the Dracula phenomenon. Within each of these groups there is considerable variation in the extent to which visitors are seeking an encounter with death or the macabre. Some literary/screen tourists may be drawn specifically to see the site of Dracula's death, although I suspect they are few in number. Most have broader motives: some are interested in seeing the places and landscapes – particularly Castle Dracula – represented in Bram Stoker's *Dracula* (and its many cinematic adaptations), while others are seeking an encounter with the Otherness of Transylvania as a wild, remote and sinister place. Their visit is underpinned by the tropes of Gothic horror, but these lead to experiences of thrill and delight rather than horror or repulsion.

A third group of Dracula tourists is those with an interest in the supernatural. Once again, this group encompasses a broad range of interests and motivations. Some Dracula tourists have personal beliefs which embrace the supernatural and a visit to Transylvania on the Dracula trail is a form of pilgrimage which offers an opportunity for a connection with the otherworldly. But for many others the supernatural is simply a source of entertainment. A visit to Transylvania offers the prospect of thrills and delight in being spooked (and even frightened) by an encounter with the supernatural (characteristics shared by some literary/screen tourists). This might appear to be underpinned by the macabre, but such engagement with the supernatural is an experience based on 'pretend' horror (in which thoughts of death may be entirely absent). For these participants such experiences are exciting rather than horrific, and for many Dracula tourists the boundaries between horror and play are very blurred. In short, the diversity of motivations and experiences among Dracula tourists in Transylvania mean that identifying it unequivocally as a form of dark tourism is problematic. It is possible to say that it may represent a form of dark tourism, for some tourists, for some of the time. This makes it the 'palest' form of dark tourism in Sharpley's formulation, involving tourists with a minimal interest in death, visiting places not intended as dark attractions.[55]

The underlying theme of this chapter has been the difficulties involved in defining what is (and is not) dark tourism. Existing typologies and classifications of dark tourism have sought to delimit dark tourism on the basis of such questions as 'what it is', 'why it is provided?' or 'why visitors go to dark sites'. However, all of these approaches have their weaknesses, and none have been totally successful in defining the nature and scope of dark tourism. For this reason, it is time to put such typologies to one side and consider dark tourism in new ways. Recent work has conceptualised dark tourism as a form of 'mortality mediation'.[56] That is, dark tourism is one of a number of ways in which contemporary societies (and individuals within those societies) engage with and negotiate (absent) death. Thus dark tourism represents a form of filter (or contact point) between life and death which can help individuals to reflect on, and understand, mortality without fear or dread. This process can obviously unfold in a multitude of ways, but the key point is that this way of thinking moves the focus away from 'what is dark tourism?' to 'what does dark tourism *do*?' This approach puts the emphasis on the cultural 'work' and meaning-making that is involved in dark tourism within contemporary societies.[57] Considered from this perspective, Dracula tourism can be used to throw light on the playful negotiation of horror in leisure settings, the role of the imagination in the tourist experience and the way in which tourist practices reproduce myths about 'Other' places. In terms of dark tourism research, this is an opportunity to move the debate on from well-worn issues such as typologies, authenticity, commodification and ethics to put questions of performance, agency, ritual, play, identity and experience at centre stage.[58]

Notes

1 M. Foley and J. J. Lennon, "JFK and Dark Tourism: A Fascination with Assassination," *International Journal of Heritage Studies* 2, no. 4 (1996): 198.
2 A. V. Seaton, "Guided by the Dark: From Thanatopsis to Thanatourism," *International Journal of Heritage Studies* 2, no. 4 (1996): 240.
3 A. V. Seaton, "Thanatourism and Its Discontents: An Appraisal of a Decade's Work with some Future Issues and Directions," in *The Sage Handbook of Tourism Studies*, ed. T. Jamal and M. Robinson (London: Sage, 2009), 523.
4 P. Stone, "A Dark Tourism Spectrum: Towards a Typology of Death and Macabre Related Tourist Sites, Attractions and Exhibitions," *Tourism: An Interdisciplinary International Journal* 54, no. 2 (2006): 146.
5 P. R. Stone, "Dark Tourism and the Cadaveric Carnival: Mediating Life and Death Narratives at Gunther von Hagens' Body Worlds," *Current Issues in Tourism* 14, no. 7 (2011): 685–701.
6 A. V. Seaton, "War and Thanatourism: Waterloo 1815–1914," *Annals of Tourism Research* 26, no. 1 (1999): 131.
7 Oxford English Dictionary, "macabre", accessed May 15, 2015, www.oed.com/view/Entry/111744.
8 These examples were taken from www.darktourism.com, accessed March 7, 2016.
9 Seaton, "Thanatourism," 525.
10 See P. R. Stone and R. Sharpley, "Deviance, Dark Tourism and 'Dark Leisure': Towards a (Re)configuration of Morality and Taboo in Secular Society," in *Contemporary Perspectives in Leisure: Meanings, Motives and Lifelong Learning*, ed. S. Elkington and S. J. Gammon (London: Routledge, 2014), 56.
11 M. S. Bowman and P. Pezzullo, "What's So 'Dark' about 'Dark Tourism'?: Death, Tours and Performance," *Tourist Studies* 9, no. 3 (2010): 191.
12 A. Biran and Y. Poria, "Reconceptualising Dark Tourism," in *The Contemporary Tourist Experiences: Concepts and Consequences*, ed. R. Sharpley and P. Stone (London: Routledge, 2012), 62.
13 Ibid., 67.

14 R. Sharpley, "Shedding Light on Dark Tourism: An Introduction," in *The Darker Side of Travel: The Theory and Practice of Dark Tourism*, ed. R. Sharpley and P. R. Stone (Bristol: Channel View, 2009), 6.
15 J. Sather-Wagstaff, *Heritage that Hurts: Tourists in the Memoryscapes of September 11* (Walnut Creek: Left Coast Press, 2011), 71–2.
16 Stone, "Dark Tourism Spectrum," 152–7.
17 P. R. Stone, "'It's a Bloody Guide': Fun, Fear and a Lighter Side of Dark Tourism at the Dungeon Visitor Attractions, UK," in *The Darker Side of Travel: The Theory and Practice of Dark Tourism*, ed. R. Sharpley and P. R. Stone (Bristol: Channel View, 2009), 169.
18 R. Sharpley, "Travels to the Edge of Darkness: Towards a Typology of 'Dark Tourism'," in *Taking Tourism to the Limits: Issues, Concepts and Managerial Perspectives*, ed. M. Aicken, S. J. Page, and C. Ryan (Amsterdam: Elsevier, 2005), 224–6.
19 Ibid., 225; Sharpley, "Shedding Light," 19; Biran and Poria, "Reconceptualising," 64.
20 Diem-Trinh Thi Le and D. G. Pearce, "Segmenting Visitors to Battlefield Sites: International Visitors to the Former Demilitarized Zone in Vietnam," *Journal of Travel and Tourism Marketing* 28, no. 4 (2011): 461.
21 A. V. Seaton and J. J. Lennon, "Thanatourism in the Early 21st Century: Moral Panics, Ulterior Motives and Alterior Desires," in *New Horizons in Tourism: Strange Experiences and Stranger Practices*, ed. T. V. Singh (Wallingford: CAB International, 2004), 74–6.
22 D. Light, "Dracula Tourism in Romania: Cultural Identity and the State," *Annals of Tourism Research* 34, no. 3 (2007): 757–8.
23 D. Light, *The Dracula Dilemma: Tourism, Identity and the State in Romania* (Farnham: Ashworth, 2012), 79.
24 B. Stoker, *Dracula*, ed. N. Auerbach and D. Skal (New York: Norton, 1997 [1897]), 10.
25 L. Brown, "Tourism and pilgrimage: Paying homage to literary heroes", *International Journal of Tourism Research* 18 no. 2 (2016): 168–9.
26 Stone, "Dark Tourism Spectrum," 152–3.
27 J. Connell and D. Meyer, "Ballamory Revisited: An Evaluation of the Screen Tourism Destination–Tourist Nexus," *Tourism Management* 30, no. 2 (2009): 194.
28 N. Tooke and M. Baker, "Seeing is Believing: The Effect of Film on Visitor Numbers to Screened Locations," *Tourism Management* 17, no. 2 (1996): 93; M. Larson, C. Lundberg, and M. Lexhagen, "Thirsting for Vampire Tourism: Developing Pop Culture Destinations," *Journal of Destination Marketing and Management* 2, no. 2 (2013): 75.
29 E. McEvoy, "Gothic Tourism," in *The Gothic World*, ed. G. Byron and D. Townsend (London: Routledge, 2014), 476–86.
30 D. Light, "Performing Transylvania: Tourism, Fantasy and Play in a Liminal Place," *Tourist Studies* 9, no. 3 (2009): 249.
31 S. Roesch, *The Experience of Film Location Tourists* (Bristol: Channel View, 2009), 159–64; A. Buchmann, K. Moore and D. Fisher, "Experiencing Film Tourism: Authenticity and Fellowship," *Annals of Tourism Research* 37, no. 1 (2010): 238–42.
32 D. Light, "Taking Dracula on Holiday: The Presence of 'Home' in the Tourist Encounter," in *The Cultural Moment in Tourism*, ed. L. Smith, E. Waterton, and S. Watson (London: Routledge, 2012), 72.
33 Ibid., 74.
34 D. Inglis and M. Holmes, "Highland and Other Haunts: Ghosts in Scottish Tourism," *Annals of Tourism Research* 30, no. 1 (2005): 52–4, 56.
35 R. S. Bristow and M. Newman, "Myth vs. Fact: An Exploration of Fright Tourism," in K. Bricker and S. J. Millington, eds, *Proceedings of the 2004 Northeastern Recreation Research Symposium* (Newtown Square PA: USDA Forest Service, Northeastern Research Station General Technical Report NE-326: 2005), 216. Accessed March 12, 2008, www.fs.fed.us/ne/newtown_square/publications/technical_reports/pdfs/2005/326 papers/bristow326.pdf.

36 G. W. Gentry, "Walking with the Dead: The Place of Ghost Walk Tourism in Savannah, Georgia," *Southeastern Geographer* 47, no. 2 (2007): 222.
37 L. Krebbs, "Spook Tourists: We Don't Want 'em but They Keep on Coming" (paper presented at the Association of American Geographers Annual Meeting, Chicago, March 9, 2006).
38 B. D'Harlingue, "Specters of the U.S. Prison Regime: Haunting Tourism and the Penal Gaze," in *Popular Ghosts: The Haunted Spaces of Everyday Culture*, ed. M. del Pillar Blanco and E. Peeren (New York: Continuum, 2010), 133.
39 Inglis and Holmes, "Ghosts," 51–3, 56.
40 Gentry, "Walking," 222; J. Holloway, "Legend Tripping in Spooky Spaces: Ghost Tourism and Infrastructures of Enchantment," *Environment and Planning D: Society and Space* 28, no. 4 (2010): 619–20; M. M. Hanks, "Re-imagining the National Past: Negotiating the Roles of Science, Religion and History in Contemporary British Ghost Tourism," in *Contested Cultural Heritage: Religion, Nationalism, Erasure and Exclusion in a Global World*, ed. H. Silverman (New York: Springer, 2011), 125; McEvoy, "Gothic Tourism," 480.
41 R. Sheehan, "Tourism and Occultism in New Orleans' Jackson Square: Contentious and Cooperative Publics," *Tourism Geographies* 14, no. 1 (2012): 81–2; B. A. W. Heidelberg, "Managing Ghosts: Exploring Local Government Involvement in Dark Tourism," *Journal of Heritage Tourism* 10, no. 1 (2014): 79–82.
42 S. Pile, *Real Cities: Modernity, Space and the Phantasmagorias of City Life* (London: Sage, 2005), 123–8.
43 K. Spracklen and B. Spracklen, "The Strange and Spooky Battle over Bats and Black Dresses: The Commodification of the Whitby Goth Weekend and the Loss of a Subculture," *Tourist Studies* 14, no. 1 (2014): 94–6.
44 Holloway, "Legend Tripping," 626, 629.
45 See Bristow and Newman, "Myth vs. Fact," 220.
46 Holloway, "Legend Tripping," 618.
47 Stone, "It's a Bloody Guide," 173.
48 D. Phillips, "Narrativised Spaces: The Functions of Story in the Theme Park," in *Leisure/Tourism Geographies*, ed. D. Crouch (London: Routledge, 1999) 101.
49 C. Laws, "Pagan Tourism and the Management of Ancient Sites in Cornwall," in *Dark Tourism and Place Identity: Managing and Interpreting Dark Places*, ed. L. White and E. Frew (Routledge: London, 2013), 98.
50 Oxford English Dictionary, "supernatural," accessed March 21, 2016, www.oed.com/view/Entry/194422
51 Sharpley, "Travels," 220.
52 R. Sharpley, "Tourism, Religion and Spirituality," in *The Sage Handbook of Tourism Studies*, ed. T. Jamal and M. Robinson (London: Sage, 2009), 241.
53 J. Digance, "Religious and Secular Pilgrimage: Journeys Redolent with Meaning," in *Tourism, Religion and Spiritual Journeys*, ed. D. J. Timothy and D. H. Olsen (London: Routledge), 36, 41–2, 46.
54 Sharpley, "Tourism, Religion," 244; Light, "Performing," 252.
55 Sharpley, "Travels," 225–6.
56 P. Stone, "Dark Tourism and Significant Other Death: Towards a Model of Mortality Mediation," *Annals of Tourism Research* 39, no. 3 (2012), 1565–87.
57 See L. Smith, "The Cultural 'Work' of Tourism," in *The Cultural Moment in Tourism*, ed. L. Smith, E. Waterton, and S. Watson (London: Routledge, 2012), 211–15.
58 See Bowman and Pezzullo, "What's So 'Dark'," 199.

10 Genocide tourism in Rwanda

Contesting the concept of the 'dark tourist'

Richard Sharpley and Mona Friedrich

Introduction

The Kigali Genocide Memorial (KGM) opened in April 2004 on the tenth anniversary of the Rwandan genocide. Constructed as a joint venture between Kigali City Council and the UK-based Aegis Trust on the site where the remains of more than 250,000 people who died in the genocide are buried, it is primarily intended as a 'permanent memorial to those who fell victim to the genocide and ... as a place for people to grieve those they lost'.[1] However, it is by no means the only such memorial in Rwanda. Across the country, a 'national network' of more than four hundred museums and memorials has been established,[2] standing in stark contrast to the vision of the country described by the national tourism authority as 'a green undulating landscape of hills, gardens and tea plantations'.[3] Often located on the site of massacres and mass graves, these memorials not only reflect survivors' determination that the atrocities of 1994 should not be forgotten but also act as a 'constant reminder of the tragic events which transpired at countless sites of violence throughout the country'.[4]

In addition to their commemorative and reconciliation purposes, Rwanda's genocide memorials have also collectively become 'dark' tourism attractions that, along with the country's national parks and other natural and cultural sites, are frequently listed in contemporary tourist guides as 'things to see'. However, since opening in 2004, the KGM in particular has arguably become the best known and certainly the most visited genocide memorial site in Rwanda. It is now an established feature of Kigali city tours and attracts significant numbers of both domestic and international visitors (see Table 10.1). Moreover, as evidenced by the personal comments in the Centre's VIP Visitor book, it is visited by most, if not all, visiting dignitaries and world leaders. For example, Tony Blair, former British Prime Minister, wrote on his visit in February 2008: 'This is moving and motivating, a reminder of the evil that humanity can do, a command that in the future it does good.'

The KGM is, then, an established tourist destination and cultural site, commemorating a tragic event that undoubtedly defines contemporary Rwanda. It is also one of a number of so-called 'genocide tourism' destinations – that is, sites of or associated with genocide – around the world that are attracting tourists in increasing numbers.[5] For example, Auschwitz-Birkenau concentration camp in

Table 10.1 Visitors to Kigali Memorial Centre

	2007	2008	2009	2010	2011	2012	2013
National	44,449	45,548	35,840	17,985	21,461	27,822	21,522
International	27,996	35,107	35,060	36,987	42,377	48,164	42,683
Total	72,445	80,655	70,900	54,972	63,838	75,986	64,205

Source: Kigali Memorial Centre (2014).

Poland, one of numerous sites associated directly or indirectly with the Holocaust, attracted a record 1.43 million visitors in 2012, while the Tuol Sleng Museum of Genocide and the Cheung Ek Genocidal Centre in Cambodia are now on the itineraries of many of the rapidly growing numbers of international tourists to that country. However, tourists' attraction to such sites remains both uncertain and contentious. More specifically, genocide memorial sites, as well as other sites memorialising death and suffering on a large scale, fulfil a variety of vital purposes, including not only remembrance but also education and, where appropriate, reconciliation.[6] However, the promotion, exploitation and commodification of genocide sites and memorials as tourist attractions is controversial both from an ethical point of view and, in particular, with respect to tourists' motivations. Indeed, there are many who regard such a form of tourism, or, more specifically, the tourists who visit genocide sites, with a critical perspective. For example, in an editorial in the *Journal of Genocide*, Dominik Schaller observes that 'it is, after all, the great demand for trips to former concentration camps and killing fields that makes genocide tourism possible in the first place', going on to claim that

> It is, not least, the entertainment value ... of rotting bodies and pyramids of skulls that make people flock to death sites ... former killing fields and concentration camps have degenerated into the ghost trains of the twenty-first century that meet the voyeuristic needs of tourists.[7]

Others similarly suggest that visiting genocide memorials and other sites of death and suffering is little more than the manifestation of voyeurism on the part of visitors, in so doing contributing to the pejorative description of them as dark tourists.[8] Indeed, the phenomenon of dark tourism more generally is often defined from a consumption perspective, as for example 'visitation to places where tragedies or historically noteworthy death has occurred' or as 'the act of travel to sites associated with death, suffering and the seemingly macabre'.[9] In both cases, dark tourism is defined as a behavioural phenomenon; that is, people visiting sites or attractions that are 'dark'. As a consequence, any tourist visiting any type of dark destination is generalised as a dark tourist and implicitly burdened with the pejorative implications of the term – as possessing and being motivated by a ghoulish, morbid or voyeuristic interest or fascination in death and suffering.

It is not surprising, therefore, that the terms dark tourism and, in particular, dark tourist are being increasingly questioned.[10] That is, it is argued that such is the diversity of the (frequently positive) motivations and experiences of visitors to dark tourism sites and attractions that to label them collectively as dark tourists is not only simplistic and inaccurate but also, given the subjective nature of the term, represents 'an impediment to detailed and circumstantial analyses of tourist sites and performances'.[11] Thus, the purpose of this chapter is to contribute to this debate. More specifically, drawing on empirical research into the experiences of tourists to the KGM in Rwanda, it seeks not only to rebuff the claim that genocide tourism simply fulfils the 'voyeuristic needs of tourists' in particular, but also to add substance to the argument that dark tourism is, at best, an unhelpful term more generally. The first task, then, is to review briefly the concept of dark tourism, before introducing genocide tourism in Rwanda as the focus of the research discussed later in the chapter.

Revisiting the concept of dark tourism

It is widely recognised that the practice of dark tourism has existed for a significantly longer period than its conceptualisation as a specific sub-field in tourism studies. As observed elsewhere, as long as people have been able to travel they have been drawn, purposefully or otherwise, towards places or events that are associated in one way or another with death, disaster and suffering.[12] However, recent years have witnessed an apparent increase in both the provision and consumption of dark tourism experiences: 'the commodification of death for popular touristic consumption, whether in the guise of memorials and museums, visitor attractions, special events and exhibitions or specific tours, has become a focus for mainstream tourism providers'.[13]

On the one hand, this growth in dark tourism might be confirmed by the relative increase in the actual number of sites and attractions (and the numbers of tourists visiting them) that fall under its broad definition. On the other, it might simply be an assumption that reflects the growing awareness and popularity of the notion of dark tourism in academic and media circles. Either way, the academic study of dark tourism, though having emerged much more recently than its practice, is also now well established. The term 'dark tourism' itself was coined almost two decades ago, although the relationship between tourism and places of death and suffering had been the focus of earlier work, such as the interpretation of war sites, while a major theme in Tunbridge and Ashworth's exploration of dissonant heritage is the interpretation of the Holocaust at the concentration camps.[14] Since then, and reflecting perhaps an alleged wider societal interest or fascination in death, the increase in academic attention paid to the concept of dark tourism has been such that it has become one of the more popular fields of study within the tourism academy.[15]

It is beyond the scope of this chapter to consider the dark tourism literature in any detail; recent reviews do so more than adequately.[16] The important point is, however, that despite the burgeoning academic attention paid to it, understanding of dark tourism remains relatively limited and 'poorly conceptualised'.[17] This reflects, in part, the fact that dark tourism is a broad umbrella term applied to a wide variety of sites, attractions and experiences, and the consequential fuzziness between it and other labels attached to the same sites, attractions and experiences, whether sub-labels (e.g. battlefield tourism, disaster tourism, genocide tourism) or alternative concepts, such as heritage, education or genealogy tourism. Equally, it reflects, in part, uncertainty with respect to the status of dark tourism as a 'product' (a place of or associated with death and suffering) or the touristic consumption of that place. Certainly, much attention was initially devoted to the identification, labelling and management of different categories of dark attractions (that is, predominantly a supply perspective), while only more recently has the significance of the consumption of dark tourism enjoyed increasing academic scrutiny.

However, the unavoidable difficulty facing the legitimisation and objectivity of dark tourism research is, as others have noted, the word 'dark' itself.[18] On the one hand, the juxtaposition of 'dark' and 'tourism' creates an appealing term that immediately draws attention (hence its use in journalistic circles) and, from an academic perspective, inspires interrogation. On the other, 'dark' or 'darkness' is considered, at least in western cultures, to connote 'something disturbing, troubling, suspicious, weird, morbid or perverse'; hence, its use in the context of tourism immediately injects a subjectively negative element.[19] It suggests, perhaps, that dark sites might be distasteful, unethical or exploitative in the way they interpret or commemorate tragic events and that the tourists who visit them, as suggested by Schaller with respect to tourism to genocide sites, possess a morbid fascination or curiosity in death, or engage in voyeurism or *schadenfreude*. This may sometimes be the case. Those who immediately rush to gaze on disaster sites, such as those who travelled to Lockerbie in Scotland following the bombing of Pan Am 103 in 1988, can only be motivated by some morbid (or deviant?) desire to witness the death or suffering of others, while, more specifically, it has been argued that 'there can be little doubt that an element of voyeurism is central to Holocaust tourism'.[20] Yet, there are clearly numerous instances where, for the tourist, an interest in death may be minimal or non-existent, or the association with death may be of little relevance. Raine, for example, in her study of visitors to burial grounds, identifies a continuum of purposes, from 'devotion' (mourning/pilgrims), through 'experience' (morbid curiosity) and 'discovery' (information seekers/hobbyists) to 'incidental' (sightseers/recreationists).[21]

Two conclusions may be drawn from this brief discussion. First, dark tourism should be thought of not as a category of tourism site or attraction, nor as a specific form of tourism consumption. Rather, it should be seen as a context for exploring the relationship between the tourist and the (dark) site and, hence, for exploring how the tourist understands or confronts the death and suffering that the site signifies, represents or memorialises. And second, applying the adjective 'dark' to tourists who visit such sites is inappropriate, not only because it disguises the multitude of reasons or motives for visiting numerous different categories of dark sites but also because, as the research discussed below reveals, far from being drawn by a ghoulish, voyeuristic fascination, tourists who visit the KGM do so with both trepidation and positive intent.

Genocide tourism: the Rwandan genocide and the Kigali Memorial Centre

In his editorial referred to earlier, Schaller expresses surprise that an organised form of tourism based on genocide should exist – 'the idea seemed just too bizarre and macabre to be true' – while others have observed that the pairing of the words 'tourism' and 'genocide' may seem unlikely.[22] Nevertheless, not only have many genocide sites, particularly those related to the Holocaust, become popular tourist attractions, but also increasing attention has been paid to what has been referred to as genocide tourism. Inevitably, perhaps, the Holocaust in general, and Auschwitz-Birkenau in particular, have proved to be the most fruitful focus of research into genocide tourism, although sites and memorials related to other genocides have increasingly attracted academic scrutiny. For example, tourism to sites related to the Balkan conflict and those commemorating the death of over two million Cambodians during the four-year Khmer Rouge period has been explored in the literature, although it is important to note that the latter is not formally recognised as genocide.[23] As a term, genocide was first coined by the Polish–Jewish lawyer Raphael Lemkin in 1944 to describe the systematic destruction of European Jews, and in 1948 it was subsequently adopted and formally recognised as an international crime by the United Nations Convention on the Prevention and Punishment of the Crime of Genocide (CPPCGG). Article 2 of this Convention defines genocide as 'acts committed with intent to destroy, in whole or in part, a national, ethnical, racial or religious group'. This definition remains contentious, however, not least because it is not applied to the mass killing of people on either social or political grounds. Perhaps as a consequence, the Convention has been applied in only two cases since 1948: the Rwandan genocide in 1994 and the Srebrenica massacre of 1995. Both of these have, in the context of tourism, attracted less academic attention to date than Holocaust-related sites, although there is evidence of increasing research into tourism to Rwandan genocide memorial sites.[24]

The events leading up to the Rwandan genocide are complex and remain contested, although it is widely regarded as the outcome of an anti-Tutsi genocidal ideology fostered over some two decades by the Habyarimana regime among the majority Hutu population.[25] What is certain, however, is that it was triggered by the death of President Habyarimana when, on the evening of 6 April 1994, the aeroplane in which he was travelling was shot down as it approached Kigali airport. The subsequent violence lasted 100 days and although the precise number remains unknown, more than a million people lost their lives. The majority of victims were Tutsis who, at that time, collectively represented just 14 per cent of the country's population of seven million, while the perpetrators were mostly Hutus. The violence came to an end only when the Tutsi-led Rwandan Patriotic Front (RPF), led by the country's current president, Paul Kagame, entered Kigali and established a multi-ethnic government. As a result of the genocide, approximately 75 per cent of the Tutsi minority had been killed, more than 300,000 children were orphaned and some two million Hutus, fearing retaliation, fled into neighbouring countries.[26]

Following the end of the violence, numerous memorial sites were established around the country, usually on the sites of mass graves or mass killings. Of these, four have been proposed by the Rwandan authorities for UNESCO World Heritage Site status: the KGM and the Nyamata, Murambi and Bisesero memorials.[27] They are also among those memorials most commonly visited by international tourists, being significant for both the nature and representation of the atrocities they commemorate. However, as already noted, the KGM is not only the principal memorial to the genocide in Rwanda but also the most visited. Moreover, it is arguably the least 'difficult' of the four main memorials, in as much as human remains are not on graphic display as at the other three. Nevertheless, through various forms of textual and photographic interpretation, including a display documenting the genocide, a children's memorial, an exhibition on the history of genocide worldwide and a wall of names, as well as the starkness of concrete-covered mass graves, the KGM offers a visually powerful and morally challenging memorial to the victims. Significantly, it also focuses on education: 'One of the principal reasons for the Centre's existence is to provide educational facilities. These are for a younger generation of Rwandan children, some of whom may not remember the genocide, but whose lives are profoundly affected by it.' Hence, the KGM is often visited by groups of school children and recently acted as focus for the commemoration of the twentieth anniversary of the genocide in April 2014 (see Figure 10.1).

As suggested in the introduction to this chapter, there is limited understanding of the experience of tourists to genocide sites. In other words, relatively little

Figure 10.1 Rwandan schoolchildren at the Kigali Genocide Memorial.
Photo copyright R. Sharpley.

research has been undertaken into why and how tourists visit sites of or commemorating mass death and, perhaps as a consequence, there persists the widely held perception that genocide tourism is little more than the manifestation of voyeuristic tendencies on the part of tourists. In order to address this lack of knowledge – and to challenge unsubstantiated assumptions with regard to the motives or intent of visitors to such 'dark' sites – three studies have been undertaken by the authors into tourists' experiences of genocide sites in Rwanda. The first was an exploratory study that, based on a content analysis of thirty-five travel blogs, explored international tourists' reported experiences of four genocide memorial sites around the country. This identified how tourists – frequently through emotionally charged narratives – responded to the sites and, in particular, to the graphic and shocking ways in which the genocide is represented.[28] However, as an analysis of travel blogs, this research was unable to elicit a more nuanced understanding of tourists' motives, expectations and responses. Hence, two further studies were undertaken focusing specifically on visitors to the KGM and their motives and experiences of the Memorial, the first of which sought to identify their attitudes and responses to a number of issues highlighted in the initial study. This was then followed up with an open-ended questionnaire survey which addressed tourists' experiences of the KGM more generally. As will be seen, these studies collectively reveal powerful, meaningful, sometimes disturbing, but undeniably positive experiences on the part of visitors to the Memorial which clearly contradict critics' assertions that dark/genocide tourism is primarily driven by voyeuristic or deviant intent.

Tourists' reported experiences of genocide sites

As noted above, the initial study of international tourists' experiences of genocide sites in Rwanda drew on a content analysis of travel blogs posted by tourists who had visited one or more of the sites. A total of thirty-five blogs were analysed, in which tourists describe their experience of the KGM; the churches at Nyamata and Ntarama, where thousands of people sought refuge but were slaughtered and where the skulls, bones and clothing of victims are on display; and Murambi, a former technical school where more than 40,000 Tutsis were massacred and where the preserved bodies of some 800 victims, men, women and children, are laid out on tables in so-called burial rooms for visitors to gaze upon.

From the experiences reported in the travel blogs, a number of themes emerged.

First, although the genocide sites have become an integral element of many tours to Rwanda, the majority of travel blogs implicitly revealed a positive desire on the part of tourists to experience the genocide sites to learn, rather than visiting them simply 'because they are there'. Indeed, many wished to understand, to assuage guilt (that the rest of the world let it happen), but perhaps also to satisfy a personal need to be shocked, to be horrified, to be shaken out of complacency or to feel and share hope in humanity. One blog implied it is, in a sense, a tourist's duty to visit the sites: for one tourist, 'somehow you can't (and shouldn't) forget what happened here in 1994', while for another 'there are tourist attraction sites such as the Kigali Memorial Centre that is a must-visit for insights into the worst genocide in history'.

Second, tourists' accounts of their visits to the genocide sites were both factual and emotional, with particular emphasis placed on their reactions to the graphic, uncompromising displays of victims' remains and belongings. Thus, a dominant theme was the intense feeling of shock, horror and revulsion experienced and described by most visitors to the genocide sites, not at the scale and inhumanity of the genocide, but as a response to being confronted by innumerable bodies in mass graves, the preserved corpses laid out on tables, the piles of human bones and skulls and the blood-stained clothes of victims:

> When I reached the doorway, my entire body went cold. I froze, a few steps from the entrance. I could see, along the back wall of the church, stacks upon stacks of human skulls . . . I felt a wave of nausea come over me.

Indeed, the sheer volume of skulls and bones on display had a dehumanising effect; the evident scale of the tragedy depersonalised it and the remains of any single person lost their individuality. As one blogger wrote: 'it is clear that the memorial has the intent to shock. It does. But at the same time, it is so macabre that it was hard to feel any grief when seeing the bodies.' For another, the shock value was too great: 'it didn't make me reflect on the genocide as much as it made me offended by the showing of these bodies.' Consequently, a number of visitors revealed that they experienced greater emotion when viewing victims' personal items, such as clothing or children's schoolbooks: 'having the colourful but mouldy cloth of the victims hang over their bones increased the intensity of the grief and despair we felt for these people we had never met.' More specifically, the images and stories of young victims on display in the KGM's children's gallery had the most powerful effect on tourists. In addition, the experience of the sites left many visitors feeling unable to comprehend the genocide; for one, 'I felt like I didn't have the right to understand what I had seen, because there were stories that were being told that were not mine – and never will be mine', thus hinting at an inherent dilemma in genocide tourism more generally: while the victims of genocide may benefit from having their story told, what right do outsiders have to share that story with the victims? Nevertheless, many suggested that their visit to the genocide sites was a positive experience inasmuch as it engendered a feeling of hope: 'I have never felt such shame and anger . . . Yet, seeing how far the Rwandan people have come, I have never felt so much hope.'

Overall, then, the initial research demonstrated that, far from being motivated by a voyeuristic desire to gaze on the tragedy of the genocide, not only do tourists visit the Rwandan sites for more positive reasons, but also their responses to what they encounter point to a deeper emotional engagement with the sites. Nevertheless, much of the narrative in the travel blogs focused (perhaps unsurprisingly) on tourists' responses to the actual sites and, in particular, the often stark and shocking displays of human remains with which they were confronted. Conversely, less was revealed about why and how they engaged with and responded to the event represented by the sites; that is, why they felt compelled to visit the sites and the extent to which they were able to confront not the victims' remains but the cause and manner of their violent death. Hence, two further studies were undertaken among international visitors to the KGM in order to address some of these unanswered questions.

Tourists' experiences of the Kigali Genocide Memorial

The KGM was selected as an appropriate site for the two studies not only because it is the principal and most visited memorial to the genocide but also because it offers a more diverse, yet less morally challenging visitor experience than other genocide sites in Rwanda. In other words, although the KGM is located on the site of mass graves, visitors are not confronted with the bones, preserved remains or personal belongings of victims, as at many other sites. Rather, the visitor experience is constructed around the three permanent displays referred to above: a self- (audio) guided tour around the story of the 1994 genocide, the history of genocide worldwide and the children's memorial. Thus, visitors are able to contemplate the genocide within a moral/ethical 'comfort zone' through reading narratives and gazing on (still often shocking) images, focusing their thoughts on the genocide as a whole rather than on their reactions to graphic collections of human remains. Moreover, the KGM is, implicitly, as much about the present and future of Rwanda as it is about the events of 1994.

The first of the two studies, undertaken in 2013, was based upon a self-completion attitudinal questionnaire distributed to randomly selected international visitors to the KGM. It was constructed primarily around a 22-item, 5-point Likert scale addressing a number of themes, including reasons for visiting the KGM, experiences of the KGM, knowledge and understanding of the genocide and reflections on the visit. In order to elicit more detailed responses, the second study, in 2014, asked visitors to the KGM to respond to a number of open-ended questions addressing similar issues, such as reasons for visiting, their overall experience of it, the importance of the KGM and the message that they took home from their visit. Thus, for both convenience and brevity, the key themes emerging from each study are combined in the following discussion.

With regard to the respondents, all were international visitors. Overall, approximately half were on holiday in Rwanda while the remainder had more specific purposes, including business, education and missionary work. Interestingly, relatively few holiday tourists stated explicitly that they had primarily come to Rwanda for the 'traditional' activity of seeing mountain gorillas; conversely, almost half of all respondents in the first study claimed that one of the main reasons for visiting Rwanda was to enhance their knowledge and understanding of the genocide in general, while a smaller proportion of respondents in the second study identified the KGM in particular as their reason for visiting Rwanda. Moreover, around half of respondents indicated that they either had intended or would like to visit other genocide memorials in the country, while in the second study, all respondents suggested that they would recommend a visit to the KGM to friends and relatives. It would appear, then, that the 1994 genocide and its commemoration have not only become a principal 'attraction' of Rwanda, but also that visiting at least one of the country's genocide memorial sites (typically the KGM) has become fundamental or essential to experiencing and engaging with contemporary Rwanda. Indeed, as now discussed, the key themes emerging from the two studies at the KGM reveal positive motives and outcomes that challenge pejorative assumptions with respect to the consumption of dark/genocide tourism experiences.

In their seminal text, John Lennon and Malcom Foley neatly sidestep the question of tourists' motives for visiting dark sites by suggesting that such visits are serendipitous.[29] In these studies, a small number of respondents similarly indicated a lack of prior intent on their part, that they had visited the KGM either because it was included in an organised tour or simply 'because it was there'. However, the overwhelming majority of respondents revealed that had made a positive decision to visit the memorial, the principal reason being to learn about the genocide, to understand its causes, to find out the 'truth' or 'what really happened'. Many also wanted to pay their respects to the victims, to commemorate the deaths of so many. As one respondent summarised, they had visited the KGM 'to get a clearer idea of what had happened and to pay respect to this beautiful country and people'. However, for one tourist visiting the KGM was simply 'mandatory', while for another 'It did not feel right, somehow, to come to Kigali and not see the memorial, considering the tragedy that befell the country in 1994'. Thus, the research reveals a positive picture of tourists proactively deciding to visit the KGM primarily to learn about the genocide, while, as also noted in the initial study of travel blogs, visiting the memorial is considered by some to be an imperative, a duty, for all tourists in Rwanda.

Significantly, however, although many respondents indicated that they had learned much about the genocide, the outcomes of the first study suggest that, for the most part, they still did not understand it. Specifically, most agreed that prior to their visit they had limited or no knowledge of how or why the genocide occurred and that, although they now knew what had happened, the nature and human scale of the genocide remained too large to understand. In this respect, perhaps, the KGM does not provide the answers that visitors seek, although many respondents in fact felt that they were 'intruding on the tragedy of the genocide'. In other words, the Rwandan genocide was a Rwandan tragedy for which outsiders have no right of understanding or explanation, pointing to an inherent contradiction with respect to the KGM as a tourist attraction. Tourists are drawn there for a variety of reasons and, indeed, the KGM seeks to convey the message of the genocide to as wide an audience as possible; yet, once they have experienced the memorial, tourists feel that they have intruded, that perhaps they should not have been there at all.

As noted earlier, the initial study focused primarily on tourists' reactions to the often horrific and challenging ways in which some genocide sites are presented as memorials, particularly where large quantities of human remains are on display. In these studies, respondents were asked, in effect, to reveal how the 'story' of the genocide as told by the exhibitions/displays impacted upon them. For some, that story is shocking:

> Too shocked. I would not have believed that human beings can be this brutal to their own brothers and sisters. Can anyone do this/ Are we human? Even

animals will not behave this way. I would like to know how the killers think about what they did twenty years ago...

However, many found their visit to the KGM to be a powerfully emotional experience, variously describing it as 'incredibly moving', 'profound', 'haunting and poignant', 'humbling' or 'emotionally paralysing', while one respondent described the KGM as 'emotional, honest, raw' and their experience as 'thought-provoking, difficult but moving'. Others experienced an overwhelming sense of 'sadness of what mankind is capable of', but tinged with 'anger, ashamed that the rest of the world just stood by'. For some, the experience of the KGM encouraged them to reflect on their own lives, one respondent feeling 'lucky and blessed to be alive'. Interestingly, in the initial travel blog study summarised above, of all the displays/forms of interpretation at genocide sites in Rwanda it was the pictures and short stories of individual children in the 'Children's Gallery' at the KGM that had impacted most powerfully on those writing about their experiences. The power of these images and stories was unequivocally confirmed by the two subsequent studies, with almost all respondents either agreeing with the statement that *it is the individual stories of children that most convey the horror of the genocide* or indicating in open answers that the most difficult experience at the KGM was the Children's Gallery.

Both studies sought to elicit tourists' opinions with respect to the role of the KGM in commemorating the victims of the genocide, in raising awareness and understanding of genocide in general and in promoting peace and reconciliation in Rwanda in particular. In the first of the two studies, all respondents agreed that *all tourists should visit the Kigali Genocide Memorial while in Rwanda*, and that they should do so because the genocide is considered to be *part of Rwanda's modern history*. Similarly, most agreed that the Memorial plays an important role in promoting peace and reconciliation, though some ambivalence was in evidence, as there was in the second study. That is, a number of respondents chose not to reflect on whether the KGM contributes to reconciliation and peace-building, while others simply indicated their hope that it would. However, many emphasised their belief that only through informing and educating people about the causes and horror of the genocide could it be prevented from occurring again: 'It is difficult to imagine how these things happened. As future generations grow, they need to be taught so they believe and remember and prevent it happening again.'

In the context of this chapter, perhaps the most significant outcomes of the research relate to tourists' reflections on their visit. Was it a positive experience? What would they take away with them from the experience of visiting the KGM? In the first of the two studies, all respondents indicated not only that their experience of the KGM had helped them to understand contemporary Rwanda and how the genocide had shaped the country as it is today, but also that they were pleased they had had the opportunity to visit the Memorial. Moreover, reflecting some narratives in the initial study, the great majority all strongly agreed that, following their visit to the KGM, their main feeling was *one of hope for the future of Rwanda*.

In the second study, respondents pointed to different messages they would take away with them, some mirroring the outcomes of the preceding studies, others being affected in different ways. On the one hand, for example, a number of respondents expressed either implicitly or explicitly their surprise and joy at Rwanda's recovery from the genocide and the message of hope it conveys both for Rwandans and for humanity more generally: 'Rwanda is a great lesson [of hope] for this world.' On the other, many respondents simply wrote the two words used to commemorate other major conflicts: 'never again', while some reflected on the potential for genocide to occur anywhere: 'No-one is incapable of participating in and encouraging genocide' and 'As individuals, we have a responsibility to pursue love, justice, truth and compassion – the human capacity for evil is so close and real in all of us'.

From the evidence of the research presented in this chapter, then, it is clear that visitors to genocide memorial sites in Rwanda, specifically the KGM, cannot be described as voyeurs. On the contrary, the three studies summarised here suggest collectively that, overall, visits to Rwandan genocide sites are positive in both intent and outcome. That is, not only do the majority of tourists visit for positive reasons of learning, understanding and commemoration, but also they recognise the positive role of the KGM in reconciliation and education and reflect on the necessarily powerful messages conveyed by the Memorial, not only to themselves as individuals but also for the future of Rwanda and humanity more generally. This, in turn, adds substance to the argument that dark tourism should no longer be conceptualised as a specific tourist market or a form of tourism consumption. Certainly, it is a valid context for exploring how places of or associated with death, suffering and atrocity can mediate between the living and those represented or commemorated by a particular site. However, to pejoratively generalise visitors to such places as 'dark tourists' is to ignore their typically positive motives and their personal and meaningful experiences and also, perhaps, to dishonour those commemorated. In short, it is time to sever the link between the words 'dark' and 'tourist'.

Notes

1 *Jenoside: Kigali Memorial Centre*, Kigali Genocide Memorial/Aegis Trust, 2010.
2 N. Mirzoeff, "Invisible Again: Rwanda and Representation after Genocide," *African Arts* 38, no. 3 (2005): 36–9; 86–91.
3 Rwanda Development Board, "Welcome to Rwanda," accessed January 10, 2014, www.rwandatourism.com/.
4 M. Friedrich and T. Johnston, "Beauty versus Tragedy: Thanatourism and the Memorialisation of the 1994 Rwandan Genocide," *Journal of Tourism and Cultural Change* 11, no. 4 (2013): 302–20.
5 J. Beech, "Genocide Tourism," in *The Darker Side of Travel: The Theory and Practice of Dark Tourism*, ed. R. Sharpley and P. Stone (Bristol: Channel View, 2009), 207–23.
6 E. Cohen, "Educational Dark Tourism at an In-populo site: The Holocaust Museum in Jerusalem," *Annals of Tourism Research* 38, no. 1 (2011): 193–209; P. Williams, *Memorial Museums: The Global Rush to Commemorate Atrocities* (Oxford: Berg, 2007).
7 D. Schaller, "From the Editors: Genocide Tourism; Educational Value or Voyeurism?" *Journal of Genocide Research* 9, no. 4 (2007): 515.
8 D. Buda and A. McIntosh, "Dark Tourism and Voyeurism: Tourist Arrested for "Spying" in Iran," *International Journal of Culture, Tourism and Hospitality Research* 7, no. 3 (2013): 214–26; D. Lisle, "Gazing at Ground Zero: Tourism, Voyeurism and Spectacle," *Journal for Cultural Research* 8, no. 1 (2004): 3–21.

9 P. Tarlow, "Dark Tourism: The Appealing 'Dark' Side of Tourism and More," in *Niche Tourism: Contemporary Issues, Trends and Cases*, ed. M. Novelli (Oxford: Elsevier, 2005), 48; A. Biran and Y. Poria, "Reconceptualising Dark Tourism," in *The Contemporary Tourist Experience: Concepts and Consequences*, ed. R. Sharpley and P. Stone (Abingdon: Routledge, 2012), 57–70.
10 M. Bowman and P. Pezzullo, "What's So 'Dark' about 'Dark Tourism'? Death, Tours, and Performance," in *Tourist Studies* 9, no. 3 (2009): 199.
11 R. Sharpley. "Shedding Light on Dark Tourism," in Sharpley and Stone, *Darker Side of Travel*, 3–22.
12 P. Stone, "Dark Tourism Scholarship: A Critical Review," *International Journal of Culture, Tourism and Hospitality Research* 7, no. 3 (2013), 307.
13 M. Foley and J. Lennon, "JFK and Dark Tourism: A Fascination with Assassination," *International Journal of Heritage Studies* 2, no. 4 (1996): 198–211.
14 Foley and Lennon, "JFK and Dark Tourism"; D. Uzzell, "The Hot Interpretation of War and Conflict," in *Heritage Interpretation, Vol. I. The Natural and Built Environment*, ed. D. Uzell (London: Belhaven Press, 1989), 33–47; J. Tunbridge and G. Ashworth, *Dissonant Heritage: Managing the Past as a Resource in Conflict* (Chichester: John Wiley, 1996).
15 G. Howarth, *Death and Dying: A Sociological Introduction* (Cambridge: Polity Press, 2007); T. Walter, "Dark Tourism: Mediating between the Dead and the Living," in Sharpley and Stone, *Darker Side of Travel*, 39–55.
16 P. Stone, "Dark Tourism Scholarship."
17 T. Jamal and L. Lelo, "Exploring the Conceptual and Analytical Framing of Dark Tourism: From Darkness to Intentionality," in *Tourist Experience: Contemporary Perspectives*, ed. R. Sharpley and P. Stone (Abingdon: Routledge, 2011), 31.
18 Biran and Poria, "Reconceptualising Dark Tourism".
19 Bowman and Pezzullo, "What's So 'Dark' about 'Dark Tourism'"? 190.
20 T. Cole, *Selling the Holocaust. From Auschwitz to Schindler: How History is Bought, Packaged and Sold* (New York: Routledge, 1999).
21 R. Raine, "A Dark Tourist Spectrum," *International Journal of Culture, Tourism and Hospitality Research* 7, no. 3 (2013): 242–56.
22 Schaller, "From the Editors": 513; Beech, "Genocide Tourism."
23 Beech, "Genocide Tourism," in Sharpley and Stone, *Darker Side of Travel*; T. Johnston, "Thanatourism and Commodification of Space in Post-war Croatia and Bosnia," in Sharpley and Stone, *Darker Side of Travel*, 43–55; O. Simic, "A Tour to the Site of Genocide: Mothers, Bones and Borders," *Journal of International Women's Studies* 9, no. 3 (2008): 320–30.
24 R. Alluri, *The Role of Tourism in Post-Conflict Peacebuilding in Rwanda* (Working paper, Bern, Switzerland: Swisspeace, 2009).
25 M. Grosspietsch, "Perceived and Projected Images of Rwanda: Visitor and International Tour Operator Perspectives," *Tourism Management* 27, no. 2 (2006): 225–34; P. Hohenhaus, "Commemorating and Commodifying the Rwandan Genocide: Memorial Sites in a Politically Difficult Context," in *Dark Tourism and Place Identity*, ed. E. Frew and L. White (Abingdon: Routledge, 2013), 142–56.
26 L. Melvern, *A People Betrayed: The Role of the West in Rwanda's Genocide* (London: Zed Books, 2009); G. Prunier, *The Rwandan Crisis: History of a Genocide*, 2nd ed. (London: Hurst & Co., 2008).
27 J. de la Croix Tabaro, 2012, "Rwanda: Genocide memorials proposed as world heritage," accessed January 22, 2014, http://focus.rw/wp/2012/07/genocide-memorials-proposed-as-world-heritage
28 R. Sharpley, "Towards an Understanding of 'Genocide Tourism': An Analysis of Visitors' Accounts of their Experience of Recent Genocide Sites," in Sharpley and Stone, *Contemporary Tourist Experience*, 95–109.
29 J. Lennon and M. Foley, *Dark Tourism: The Attraction of Death and Disaster* (London: Continuum, 2000).

11 Everyday darkness and catastrophic events[1]

Riding Nepal's buses through peace, war, and an earthquake

Sharon J. Hepburn

Although there are better and worse deaths, everyone dies. And although we all experience different kinds and degrees of suffering, no one dies without their share of it. This is everyday darkness. The very idea of dark tourism is premised on selective attention to instances of events that are, in fact, ubiquitous. In this volume Tunbridge argues that any heritage site has multiple meanings, and can be variously dark to different people, and I have previously argued that in a place where there is abject death, fear, and suffering, this can be variously seen and experienced by differently situated people.[2] I suggest here that we think of layers of darkness which are historical, and I evoke the idea of darkness accruing over time, much as sedimentation leads to geological layers. These layers of darkness usually leave no trace discernible to someone passing through. Perhaps those traces are buried in the memory of the people affected, perhaps themselves now dead. Although this might sound grim, in Nepal one does hear the word *dukkha* (suffering) far more often than its counterpart *sukkha* (happiness, or more realistically, contentment with the lack of obvious, intrusive suffering). And the heart of the philosophy of the Buddha, Nepal's most famous son, is that suffering exists as a fundamental feature of human life. Certainly, the geological analogy is just an analogy, but it is useful as tourists do travel through social landscapes built through time and memory. And although they travel to see the landscape, their journeys—on buses, the focus of this chapter—take them through landscapes marked by *dukkha*. The highways are marked by the places (and mangled vehicles) of the death of previous travelers, by the death, maiming, and destruction of a civil war from 1996 to 2006, and now by an earthquake and its aftershocks in Spring 2015.

In this chapter I describe how tourists variously brush up against these sources of darkness, are confronted by it, or are engulfed by it. Also I describe how, on the other hand, they can pass through it unknowing and oblivious. The tourists that are aware of the darkness can also variously experience these encounters: they can be fun, exhilarating, terrifying, uncomfortable, and sometimes profoundly disturbing. I conclude by reflecting on heritage sites and memorials in the midst of this everyday darkness. Although the layers of darkness rarely leave a trace discernible to a tourist passing through, sometimes they are marked by memorials and dark heritage sites. But which traces are marked is highly selective. The "light in the dark places" that Dunkley encourages in this volume—a change in the tourists' awareness of injustice—may also come from tourists' more mundane yet embodied activities, in addition to the particular crafting of sites which Dunkley advocates.

Using intrinsically appropriate mobile methods,[3] this chapter itself travels over three decades, on tourist and local buses in Nepal since I first lived there in 1986, and over roads blocked by post-earthquake debris in 2015. Its route follows tourists as they fear for their lives, see evidence of the death of others, pass through armed warfare—mostly either excited or annoyed—while those around them tremble, and smell the decay of debris-buried bodies while they walk quake-damaged roads.

Everyday darkness and bursting the immortality bubble

Many tourists come to Nepal to walk Himalayan trails. Getting to a trailhead often requires a bus journey. Like the trail, the bus passes through glorious scenery. However, in books, in blogs, or in conversation with an anthropologist such as myself, description of the ordeal of travel in the bus often surpasses reflection on landscapes, except as obstacles. As they narrate their stories, the bus tourists recount an emphatically embodied experience. Sometimes this is a thrill and "part of the adventure," and the frisson of pleasure arises when the brush with darkness outweighs the thought of truly dark or even fatal consequences. More often the journey is suffused with the discomfort and fear that comes with awareness of real danger. As Lim says of tourists at the end of a road journey to reach the start of the Langtang trek:

> almost without fail, most tourists will express relief when they reach (the trailhead) . . . regardless of the mode of transport . . . in the opinion of many . . . it is "one of the worst rides in the world" . . . conditions . . . constitute the most hazardous feature of the journey . . . the narrow dirt road perched precariously along sheer cliffs . . . Foreigners are only too happy to reach the destination without any mishap, and generally did not look forward to the return bus journey.[4]

Below we will hear more of why tourists who "survived" (they stress) on this and other roads say these things.

In beginning to ride Nepali local buses in 1986, I moved from a tourist bubble of the sort that maintains distance from local people which Cohen, Smith, and others have written about, but also the sort that allows you to believe that things can carry on as if you were at home: that authorities and institutions are protecting you with road and vehicular safety standards that they can enforce.[5] I also moved from the sort of bubble that comes from knowing you have medical and evacuation insurance, and, perhaps more importantly, an embassy to serve you. The bubble bursts as you realize you are just another human body subject to the same conditions as other people on the bus. And those are conditions of discomfort—outright suffering, from the point of view of many tourist travelers—and, indeed, mortal danger. The signs are all around, and clear to see, bursting the immortality bubble.

On their bus journeys, tourists encounter objectively dangerous conditions in a country with a very high rate of road fatalities. They are confronted with evidence that people doing exactly what they were doing have died with hardly a moment's notice. Stone notes that ordinary death in the West is sequestered behind medical and professional facades, and that dark tourism can disrupt that filtering of death.[6] In Nepal the vehicular evidence of the "ordinary" traffic death is left abandoned for all to see, not towed away. The tourists face conditions in terms of vehicular regulation, road maintenance, driving practices, and comfort very different from those at home. Although they are not visiting a dark tourism site, they still move through a landscape marked by the deaths of many people. They are not visiting a site, at some sort of physical or historical remove from the darkness; they are brushing up against past and potential accidental death, or what I am here calling everyday darkness.

The objective dangers of bus travel in Nepal

Road travel in Nepal of any kind is dangerous relative to most other countries. The WHO reports that in Nepal there are 1,677 fatalities per 100,000 vehicles per year; in comparison India has just 100, and China 36.[7] The media covers the worst accidents. Some of these involve tourists: for example, in October 2014, two Israeli tourists heading to trek in the Langtang district died when their bus went over a cliff, and in March 2015 four Korean tourists died when their minibus was overtaking and hit a larger bus coming towards them in dense fog.[8] In April 2015, a few days before a 7.8 earthquake shook Nepal, a bus went off a narrow road at night, killing 17 pilgrim tourists.[9] It is well known that the accidents are caused by the condition of poorly maintained, slippery, narrow, mountain roads; landslides (that have to be crossed); overcrowding; carelessness; drivers who are drunk or reckless or both; negligence; and irregular maintenance of vehicles.[10] T. B. grimly titles his article in *The Economist* "Double-O driver" (i.e. license to kill), and cites police officials in Nepal saying that the "syndicate system, a cartel ... is very powerful" and makes enforcing laws very difficult.[11] As tourists travel the highway, they pass the sites of these crashes and plunges.

For four months in 1986, while working in rural Nepal, I took a weekly two-hour journey on a local and frequently packed bus. I have ridden local buses for longer journeys. I have also traveled on much less densely packed designated tourist buses which are overall in better condition, more expensive, and have assigned seats, although obviously they encounter the same road conditions as local buses. During those years I have listened to tourists on buses, and specifically interviewed more than fifty about their bus journeys in the general context of talking about their travels. About half of the interviews used here were in the early 1990s, and the other half during the civil war between the Communist Party of Nepal-Maoist and the armed forces of Nepal, when non-transport related forms of incipient and past darkness hovered. The rest of this chapter is based on those interviews and encounters, and also on blogs, online forum discussions, and paper-published travel accounts.

Local buses and many tourist buses in Nepal are generally dilapidated. I have heard tourists note that buses are "on their last legs," "ones India no longer wants to use," "cages on wheels," or "death traps." Some tourists mention specifics such as "no suspension" or the bus having "no working headlights." Unsurprisingly, repair, breakdowns, and their potential dangers and pleasures are recurring themes in blogs. Some, like Brian, delightfully recount a series of breakdowns on the journey; he makes light of his concern for safety, for example, and praises the expertise of the drivers in keeping buses moving with the skills of a *bricoleur*. One driver in particular managed to get a bus with a broken transmission moving by having a helper hold a pole through a hole in the floor, thereby keeping something in the transmission system nominally in place as they traveled curving mountain roads.[12] TO, a long-term volunteer resident, relates the familiar series of stalling, false starts, and repairs on hilly roads until "the engine rumbled to life again, and we were on our way . . . As the little bus . . . continued lugging itself up the hill, it began to spew smoke out the sides" and ground to a halt.[13] Although Brian delights in the potential darkness of the journey, and TO sees more delay than danger, others such as Mary however react with fear and concern when they see the poor mechanics: She told me that she was worried during daylight about the decrepit bus she took, but as it got dark she was even more alarmed to find that it had no working headlights.

The journey is undeniably embodied, not just a facilitation of "gazing," and somatic responses to the journey are often, like Mary's, literally and inextricably bound with fear. True, again, there are tourists who choose to travel on the bus roof, and find the ride a great adventure and exhilarating, despite or in part because of the risk (as especially young tourists often do).[14] They describe in glowing terms the closeness that grows with fellow roof riders as they cling together and duck overhead wires and trees. But there are also those who choose to ride the roof for fun and quickly regret their choice, as well as those who choose the roof because they believe it gives them a better chance of survival if the bus rolls, because they can jump.[15]

Tourists have told me at the end of a bus journey their nerves have been "shredded," or that they have marks on their legs from people grabbing them tightly. One noted he might have "collapsed from the strain of holding . . . muscles taut for hours." Some have urine- (or worse) soaked pants as a fear response. Paul from England said he would never again complain about his rail commute in Britain (where he experienced what Bissell calls vibrating materialities, a comparatively soothing experience, even in the worst instances).[16]

Dave, for example, described a journey between Pokhara and Kathmandu, a route thousands of tourists travel every year. There are many tourist buses, but he and his companion took the local bus. He notes in his blog that the driver drove fast as they "soared" and "climbed rapidly" around many blind corners. He describes looking out the window, but not to "gaze" at the view. He said that, being near the front, he could see and brace for the onslaught of rapidly approaching trucks and buses—a danger to both tourist and local buses—and he suggests that the game of "chicken" was invented in Nepal. Looking to the side, he saw alarmingly precipitous gorges and hillsides and noticed a truck below on its side, and another which had landed on its roof. He tells how the bus unexpectedly stopped and the driver walked over to the road edge—"jumping excitedly, waving his arms and pointing," wanting the passengers to get out and look over the edge where he pointed to the bottom of the gorge—and said "Tuesday's bus," laughing "louder now, almost maniacally." When Dave told the story later, he was laughing at events "that only hours before had us secretly praying for reprieve." Dave also "realized the terrifying truth: somewhere in Nepal, there was an abandoned bus named after every day of the week."[17] Of course, if they were to be more aptly named, buses would be named for days in the year. Dave had glimpsed the consequences of the things he feared, but did not understand how deep the layers of dead buses and their passengers went.

It is not surprising that tourists often feel fear. Phrases like "nerve-wracking" are common when people describe their journeys. Tourists also often re-enact the journey a little when recounting the tale in person: they sway and jolt their bodies, and hold their breath—as in a tense moment—gasping, saying that they sometimes thought it might be their last breath. Although hearts can "lurch with hope" (as tourist Emily told me) when a bus finally starts moving, they more likely lurch with fear as viscerally felt emotion. Tourists tell of the obvious terror of accelerating towards sheer drops. Scott, a United States Peace Corps volunteer, agrees that although it is a great way to meet local people, bus-riding is also the "near equivalent of putting a revolver in your mouth with all chambers loaded."[18] I have heard others use the Russian roulette analogy, with one saying the result is either being scared or outright terrified, and he has felt both. The unpleasantness of a crush of people on a packed local bus becomes the fear that these are the people you may die with.

In addition to reckless driving, or even with good driving, the terrain itself can move discomfort to fear, and thoughts of mortality. As a few examples above show, many roads are cut into cliffs, often suspended without barriers above ravines and rivers, sometimes swollen with monsoon water. Blogger *sublimbedisruption* described travel on a road covered with the debris of landslides, with the bus slipping and tilting in wet mud "towards nothing but a kilometer of air and the floor of a Himalayan valley." He says this is a time when you are "quite literally confronted by a sense of your own mortality . . . (with) nothing but the grip of the worn tires on the jaded wheels of a decades old Tata bus between you and oblivion."[19] On another tilting bus, Chad describes feeling like he was about to have a heart attack as he looked out his window and down and could not see any roadway, just the river far below. Chris notes that

> we were pitched and tossed around like a trawler on a squally sea. The bus lurched from one side to the other as the tyres struggled for grip in the glutinous mud, giving those of us in the window seats a close look at the huge chasm beyond the edge of the road. What started as mild panic soon became sheer terror and my thoughts went from "that was a close one," to "you've got to be kidding," to "ohmygodwereallgoingtodie!"[20]

If tourists have a general idea of the risk when they start their journey, they often take it in stride. The excitement of being there at all overrides concern, and "it's all part of the adventure," some say. Of the drivers, they might say with a shrug, "they know what they are doing." Almost no one I talked to knew of the high rate of vehicular death compared to other countries. A few people thought that because Nepal depends on tourism revenue, and news of road deaths would scare off tourists, it could not be *that* dangerous; they failed to explain that connection, except to vaguely suggest that "they" would see to it that tourists remained safe. Most, after all, are heading to see the Himalaya, and are looking forward to their first glimpse from the bus. But whatever attitude the tourists take, there emerges for some an undeniable appreciation that people doing just what they are doing have died, without warning, perhaps plunging to their deaths. Even with no memorial or dark tourism "attraction," the tourist can unexpectedly have an immersive experience in the reality of the conditions of ongoing dark—accidental, violent—deaths.

Riding Nepal's buses during the civil war, 1996–2006

Following al-Qaeda attacks upon the United States of America on September 11, 2001, tourism declined worldwide. In the case of Nepal, this external political event coincided with internal political struggles—namely a civil war and the massacre of most of the royal family—exacerbating the effects of a wider phenomenon. The *jana yudha*, or people's war, began in 1996 with the goal of replacing a multiparty system with a communist republic, a conflict that had little or no effect on tourism until 2001. Echoing the United States' call for a "war on terror," in November the Nepali government promulgated the TADO (Terrorist and Disruptive Activities Ordinance) and identified the Maoists as terrorists. King Gyanendra suspended parts of the constitution which guaranteed rights against preventive detention, while citizens could henceforth be held for ninety days without charge on only the suspicion of being a Maoist and thus a terrorist.[21] With the State of Emergency and the availability of weapons from the United States in late 2001, the conflict in Nepal escalated rapidly. A peace agreement was eventually signed in 2006, but by that time 13,000 people had died as a direct result of the insurgency, thousands more were maimed, and at least 1,400 were unaccounted for.[22] The world climate of fear and caution, as well as Nepal's 2001 state of emergency and ongoing cycle of turmoil and violence, took a heavy toll on the country's tourism. Arrivals dropped from about 464,000 in 2000 to less than 216,000 in 2002[23] as many potential visitors either stayed home or went somewhere less troubled than Nepal. Indeed, for a few years Nepal topped Amnesty International lists for the number of non-judicial executions and disappearances.[24] People could and did disappear. Maiming and torture were widespread. Civilians were shot without trial on suspicion of helping either side. Everyday life for non-aligned/non-combatant citizens—the self-described "yams between two rocks"—continued as best it could. Nonetheless, varying degrees of fear and uncertainty were the background to life. Many feared the Maoists.

Overall, however, people were more scared of the army, who, as one friend told me from her village experience (and Pettigrew confirms),[25] "just come in shooting." After 2001 there was a marked army presence in towns, and an unpredictable presence in rural areas. They set up checkpoints on highways, and any vehicle could be stopped and searched. Maoists frequently called for *bandh* (strikes) which halted all traffic; a moving bus or taxi might be burned and the driver killed. So, in addition to seeing the wrecks of vehicle accidents, tourists started to see burned-out buses along the road. Maoists laid trees or barbed wire across the highways to stop movement. Sometimes they laid bombs or landmines in the road, or threw bombs at moving buses. Even without these obstacles, all journeys took much longer than usual due to frequent army checkpoints on the road. At these, Nepalis (sometimes everyone, sometimes just men) generally had to dismount and wait while soldiers searched the bus and passengers, although tourists were generally not searched. If a bus or other vehicle had only tourist passengers, the drivers put large signs on the bus saying "Tourists Only." Soldiers waved these vehicles through checkpoints (often after long waits, however). Buses sometimes had army accompaniment, for example in Kathmandu during a *bandh*, when the army escorted buses and taxis organized by the Trekking Agents Association and the Nepal Tourism Board, and shuttled tourists between the airport and hotels.

Like me, many tourists had accessed online information and discussions and some still decided to visit Nepal despite embassy warnings to avoid non-essential travel, despite media reportage of bombings, despite Maoists asking for donations, and despite the Maoists' explicit warnings that although they did not target tourists, a tourist might get caught in the crossfire.[26] Sometimes the conflict added an exciting twist to their trip. People were often quite blind to its presence around them. The impression that "things are fine here" was carefully managed by Nepalis for many reasons, but partly for the same reasons they managed what they said to other Nepalis: you don't know who might hear and what their allegiances are.[27] One tourist described the conflict to me as a "boutique" insurgency in a picturesque setting. For him and many others, stories of the friendly Maoists coalesced with images of Shangri-la. The main problem for most tourists was inconvenience—journeys on both local and tourist buses become twice as long due to checkpoints. From 2001 until the end of the war, online discussion sites and word of mouth in Nepal conveyed tips on how to travel around various *bandhs* (general strikes) and curfews. But as the situation was unpredictable, in the end many people simply got on the bus and hoped for the best.

During some strikes and blockades, the government banned travel without a security escort on main roads, leading to long waits and long convoys. John from Ireland traveled from the southern border under such conditions, with an army presence and checkpoints unlike any he had seen anywhere. The only tourist on his bus, he noted that all the Nepalis were very tense, "because of the Maoists." Having heard that Maoists did not harm tourists, he was confident of his own safety, and he expressed no fear of the army. After ten hours on the bus, they stopped for a night curfew. John stayed on the bus all night and enjoyed the "adventure" of it. The army boarded during the night and searched everyone but

John. He had hoped for a "Maoist experience" and happily conceded that this bus journey was more (and better) than what he had anticipated. A post on Thorn Tree, an online travel discussion board, describes a similar journey in February 2005 at the start of a general strike and blockade of goods into the Kathmandu Valley. The Thorn Tree poster's bus was part of a convey of buses and petrol tankers heavily guarded by the army, including helicopter cover, and on his journey "15 soldiers jumped out of the bushes and onto the bus. They left us after an hour or so." He described his overnight on the bus as "kinda fun," acknowledging (after the prompting of another Thorn Tree poster) the relative security of his situation as a tourist which allowed that sort of experience: he said it was an adventure, unexpected, enjoyable, and unique.[28] Perhaps, like riding poorly maintained vehicles on bad roads, travel in war can be "kinda fun" if you think the risks are not so great, or if you are somehow protected by authorities and institutions that function like those at home. You know the darkness is there, but you can place yourself outside of it and savor the excitement it brings.

Designated tourist buses continued to run during the blockades and strikes, often at sharply inflated rates. Finn and Anna traveled on such a bus from Kathmandu to Pokhara. What is generally a six to seven-hour journey took them thirty-two hours. The bus traveled with an army escort along stretches of the route. About thirty kilometers from Pokhara, the road became impassible due to a large hole and tree across it. Their driver was scared that Maoists were watching nearby, ready to attack should they try to pass, so Finn and Anna ended up walking a few hours to find a local bus to complete the journey. Some tourists decided to just walk the roads during strikes. During a *bandh* in an area with no tourist buses, Steve from Canada walked the highway, around and across trees felled on the road. Hearing later that Maoists hid landmines in these trees, he felt lucky to have been spared, even though walking was supposed to be permitted by the rules of the *bandh*. He did not relish the brush with danger, and said that he had not anticipated coming "so close to the war" and actual, capricious danger; he expressed relief mixed with unease that "the rules" had been violated and the Maoists were not "being straight," and wondered if he could continue to trust that tourists would not be targeted. Likewise, Hermann and Juan traveled by bus from the Tibetan border at the end of 2004. When the driver stopped due to roadblocks and fear of bombs, they also walked and had the highway to themselves, stepping around stones and felled trees. They thought they saw a bomb on the ground hidden by branches—and clearly enjoyed the unexpected adventure of a rigorous walk on a un-trafficked road—but, like Steve, they did not expect such a direct encounter with potentially deadly conditions that were not moderated by their being tourists. Like riding the roof of a swerving bus, perhaps, the frisson of danger and darkness can be exciting so long as it does not become too close, too real.

For most tourists, flexibility and patience was key, rather than dealing with actual conflict-related danger. They needed to be willing to change route and/or destination in order to avoid curfews or strikes, to fly instead of taking a marked tourist bus (as many did, especially older tourists), and—like Finn and Anna—to be prepared to walk or wait through many checkpoints. Some found this led to unexpectedly pleasant experiences, such as walking routes rarely taken by tourists anymore rather than taking either a plane or bus. Almost all tourists said they felt safe.

Some astutely noted that really, the usual risks of the road were more worrying than the possibility of danger from Maoists. In conversations with bus travelers during the latter half of the conflict, they usually talked about how they made decisions about travel plans, and they sometimes talked about incidents around searches, check-posts and the army presence. But they almost always talked about what tourists talk about in peacetime: the rigors and dangers of the bus journey itself. The kind of accounts presented in the first part of this chapter continued to be told by tourists traveling during the war. A few tourists even noted that travel by tourist bus during the war, during a *bandh*, felt relatively safer than previous journeys, because there were fewer vehicles on the road to have accidents with. During the war, even though there may have been Maoists or military, road blocks and searches on the journey, there were still the curves, drops, drunk drivers, poorly maintained and fast-approaching vehicles, none of which, as always, paid any attention to anyone's politics.

Earthquakes and walking shattered roads, spring 2015

Mostly this chapter is about travel on buses before and during the civil war in Nepal. But it ends with another event that blocked bus and most other travel over large parts of Nepal. On April 24, 2015, almost a decade after the peace agreement was signed, the earth under Nepal heaved and jolted, and the landscape shuddered for over a month. Tourists looking for a bit of adventure found more than they bargained for. As dams broke, landslides blocked roads (more than usual), roads cracked or collapsed, buildings—including tourist hotels—fell, and in one place—the Langtang Valley—a wall of ice and debris buried a village and the people in it, including more than 100 foreigners. Tourists in Kathmandu took shelter in embassies or slept in open spaces with Nepalis. They inhaled the smell of bodies decomposing under rubble, the smell of the overworked cremation *ghats*, and feared epidemics. They sent word to friends that they were alive, not dead. Rather than wait an indeterminable time for rescue, some walked out along roads and trails, scrambling over rockfalls and debris, noting devastation all around. They saw corpses and noted the ever present stench of death along the road. "Stranded" tourists were evacuated from Nepal, and told their stories to media at home of having witnessed and escaped death, of having brushed up against darkness. A few months later tourists returned in small numbers, some to enjoy areas where facilities were undamaged and now pleasantly uncrowded and some—the dark tourists—to see the destruction and the aftermath. The Nepali government and Nepal Tourism Association were asking people to come to Nepal, as a way to help Nepal rebuild from darkness.

It was a fearsome event, widely covered by media showing damaged temples from every angle; anguished Nepalis searching through rubble; charming, resilient children; and heroic climbers and trekkers being chased by immense avalanches from mythical mountains. A natural disaster of this scale will wipe out any lingering awareness on the part of tourists that they are walking through an area that had recently been devastated by war. When the memory of the war has faded with time and the death or emigration of witnesses, other deaths and suffering will continue, including the deaths of those plunging in buses from the cleared and repaired roads.

There is an accretion of memories of death and suffering on the roads of Nepal—the layers I talked about at the start of this chapter. This is a process that goes on in any case, without war, without a 7.8 earthquake leaving people dead, homeless, and without a livelihood. And that accretion goes on continuously and obviously in this poor country, with one of the highest infant mortality rates in the world, poor sanitation, and widespread poverty (no matter how picturesque); and this death and hardship is the backdrop to any travel through Nepal.

I started by saying that tourism and heritage industries draw attention to the presence and idea of death and suffering in very particular cases. In the case of Nepal, there are already plans to build a monument to the dead in the Langtang village that was buried (organized by families of tourist victims), about which future trekkers will be curious, much as there are memorials in China from the Wenchaun 2008 earthquake.[29] Four months after the quake the town of Bhaktapur held a memorial in which they displayed photos of the hundreds of residents who died, and a permanent memorial is planned. There are also plans for memorials and plaques, and for interpretive displays of heritage structures that collapsed and the people on which they collapsed. Moreover, although the Maoist insurgency failed in that the new republic is not communist, Maoist politicians have promoted the Guerrilla Trek in the former Maoist heartlands.[30] Simpler than the Israeli conflict heritage tourism Mansfeld describes, tourists are being invited to come see the scenes of battles, the Martyrs' gates, and other residues of the conflict.[31] But few tourists have walked the trail, which is not surprising, as so few really knew the extent of the conflict. The Nepali government and foreign agencies are funding victim-centered memorial activities for the war disappeared as well as the dead,[32] but certainly, as in Cambodia,[33] the project of memorializing the war is a low priority for a struggling government, and resolution and healing a long way off for many.

At the outset of this chapter I noted that there are better and worse deaths. Dark tourism draws attention to worse deaths, such as the deaths of the Holocaust or a massacre. Heidelberg talks of how dark events and sites can leave a dark ghost "lingering in (the) . . . shadows" that local governments must manage if the site or polity is to retain a good image for potential visitors.[34] Along the roads of Nepal there are many dark ghosts lingering: of people who rode buses, or who died in the war, and now also the ghosts of those who died in the earthquake: the dark ghosts of bad deaths. Yet, if any are to be memorialized in a way with which tourists will engage, it will most likely be the earthquake victims (human or architectural). And yet in the past nine years more people have died in vehicle accidents than died in either the decade-long war or the earthquake and aftershocks.

Most of this chapter has been about the discomfort and fear and the objective dangers of bus-riding in Nepal: this is more dangerous than a war, and more persistently deadly than a 7.8 earthquake. These are the dangers of life in a very poor country. Our tourists can see the danger, and fear mortality when they are part of it; when they walk around bombs on felled trees, when they feel the earth lurch beneath them, or pass decaying corpses lying in collapsed homes. But despite the more dramatic media-covered events, more commonly they confront it when their bus travel scares them, when they are immersed in the dangers of life on the roads, on the buses. In traveling the roads of Nepal, during war, relative peace, or utter destruction with engulfing suffering, there is a window for tourists—in their brushes with darkness—to reflect on their circumstances and those of the people among whom they travel. In this volume Dunkley hopes that 'thanasites' could lead to a cultural shift required to end world injustices; in the case of Nepal, perhaps this could come from awareness of the everyday darkness in the landscapes through which they travel, rather than the darkness of catastrophic events.

Notes

1 This chapter is based on research generously funded by the following agencies: the United States National Science Foundation (Grant N. BNS-9100365), the Joint Committee on South Asia of the Social Science Research Council and the American Council of Learned Societies with funds provided by the Andrew W. Mellon Foundation, and the Social Sciences and Humanities Research Council of Canada.
2 S. Hepburn, "Shades of Darkness: Silence, Risk, and Fear among Tourists and Nepalis During Nepal's Civil War," in *Writing the Dark Side of Travel*, ed. J. Skinner (Oxford: Berghahn Books, 2012), 122–42.
3 M. Buscher and J. Urry, "Mobile Methods and the Empirical," in *European Journal of Social Theory* 12, no. 1 (2009): 99–116.
4 F. K. G. Lim, "Of Reverie and Emplacement: Spatial Imaginings and Tourism Encounters in Nepal Himalaya," in *Inter-Asia Cultural Studies* 9, no. 3 (2008): 379.
5 E. Cohen, "Towards a Sociology of International Tourism," in *Social Research* 39, no. 1 (1972): 164–82; V. Smith, "Introduction," in *Hosts and Guests: The Anthropology of Tourism*, ed. V. Smith (Philadelphia: University of Pennsylvania Press, 1978), 1–21.
6 P. Stone, "Dark Tourism and Significant Other Death: Towards a Model of Mortality Mediation," *Annals of Tourism Research* 39, no. 3 (2012): 1565–87.
7 World Health Organization, *Global Status Report on Road Safety: Supporting a Decade of Action* (Geneva, Switzerland: World Health Organization, 2013), 165.
8 "14 Dead in Nepal Bus Crash," 25 October 2014, *The Guardian*, accessed March 22, 2016, www.theguardian.com/world/2014/oct/25/14-dead-in-nepal-bus-crash; "Four South Korean Tourists Killed in Nepal Bus Crash," March 30, 2015, Zee News, accessed March 22, 2016, http://zeenews.india.com/news/world/four-south-korean-tourists-killed-in-nepal-bus-crash-police_1570347.html
9 "17 Indian Pilgrims Killed in Nepal Bus Accident," April 22, 2015, www.dailymail.co.uk, accessed March 22, 2016, www.dailymail.co.uk/wires/ap/article-3049935/12-Indian-pilgrims-killed-Nepal-bus-accident.html
10 D. A. C. Maunder and T. C. Pearce, *Bus Accidents: An Additional Burden for the Poor* (Crowthorne, Berkshire: DFID Transport Research Laboratory, 2000); D. A. C. Maunder and T. C. Pearce, "Bus Accidents in the Kingdom of Nepal: Attitudes and Causes" (Crowthorne, Berkshire: DIFD Transport Research Laboratory, 1998).
11 T. B., "Double-O Driver," *The Economist*, August 16, 2012, accessed March 22, 2016, www.economist.com/comment/1581636
12 Brian, "Anatomy of a Bus Breakdown in Nepal," June 4, 2013, accessed March 22, 2016, http://wanderingsasquatch.com/anatomy-of-a-bus-breakdown-in-nepal/

13 "Dhalbhatt, A Blog of Peace Corps Service in Nepal: Bus-ted," June 25, 2013, accessed March 22, 2016, https://daalbhaat.wordpress.com/2013/06/25/bus-ted/
14 P. Mura, "'Scary . . . But I Like It!' Young Tourists' Perceptions of Fear on Holiday," in *Journal of Tourism and Cultural Change* 8, no. 1–2 (2010), 30–49.
15 R. Gilbert and G. Davis (presenters), *Dangerous Roads: Nepal*, Series One, Part Two (BBC: 2011) www.youtube.com/watch?v=5qr7mSiUjBE
16 D. Bissell, "Vibrating Materialities: Mobility-Body-Technology Relations," *Area* 42, no. 4 (2010): 479–86.
17 D. Underwood, "Through Nepal by Bus" (n.d.), accessed March 22, 2016, www.goworldtravel.com/travel-nepal-by-bus/
18 S. A. Wallick, "Thoughts on Transportation in Nepal," *The Peace Corps Experience of Scott Allan Wallick*, September 28, 2002, accessed March 22, 2016, http://peace-corps.scott.wallick.org/blog/2002/09/28/thoughts-on-transportation-in-nepal/
19 Sublimedisruption, "Nepal: The Road to Enlightenment?" November 7, 2011, accessed March 22, 2016, http://sublimedisruption.tumblr.com/post/11347000367/nepal-the-road-to-enlightenment
20 C. Parson, "A Tale of Two Buses: A White-knuckle Ride in Nepal," Parsons on Tour, September 22, 2011, accessed March 22, 2016, http://parsonsontour.travellerspoint.com/6/
21 M. Hutt, "Introduction: Monarchy, Democracy and Maoism in Nepal," in *Himalayan People's War*, ed. M. Hutt (Bloomington: Indiana University Press, 2004), 1–20.
22 K. Bhattarai, D. Conway, and N. Shrestha, "Tourism, Terrorism and Turmoil in Nepal," *Annals of Tourism Research* 32, no. 3 (2005): 669–89.S. Shah, "Revolution and Reaction in the Himalayas: Cultural Resistance and the Maoist New Regime in Western Nepal," *American Ethnologist* 35, no. 3 (2008): 418–99; International Committee of the Red Cross, *Missing Persons in Nepal: The Right to Know* (Kathmandu: IDRC, 2012).
23 MTCA, *Nepal Tourism Statistics* (Kathmandu: Ministry of Tourism and Civil Aviation, 2003).
24 Amnesty International, *Nepal. A Deepening Human Rights Crisis: Time for International Action* (London: Amnesty International, 2002).
25 J. Pettigrew, "Living between the Maoists and the Army in Rural Nepal," in *Himalayan People's War: Nepal's Maoist Rebellion*, ed. M. Hutt (Bloomington: Indiana University Press 2004), 261–83; J. Pettigrew, *Maoists at the Hearth: Everyday Life in Nepal's Civil War* (Philadelphia: University of Pennsylvania Press, 2013).
26 The most widely read online boards during the conflict were *Thorn Tree* on www.Lonelyplanet.com and the Nepal-specific www.yetizone.com, which has now been removed from the internet; P. Bjork and H. Kauppinen-Raisanen, "A Netographic Examination of Travelers' Online Discussions of Risks," in *Tourism Management Perspectives* 2, no. 3 (2012): 65–71.
27 N. Gautam, *Maoist Movement and Development of Silence Culture: A Case Study of Dang District in Mid-Western Nepal* (Kathmandu: Tribhuvan University, 2004); Hepburn, "Shades of Darkness"; M. Lecomte-Tilouine, "Terror in a Maoist Model Village, Mid-Western Nepal," *Dialectical Anthropology* 33 (2009): 383–401; J. Pettigrew, "Learning to Be Silent: Change, Childhood and Mental Health in the Maoist Insurgency in Nepal," in *Nepalis Inside and Outside Nepal: Political and Social Transformations*, ed. I. Hiroshi, D. N. Gellner, and K. Nawa (Delhi: Manohar, 2007): 307–47.
28 www.lonelyplanet.com/thorntree/forums/asia-indian-subcontinent
29 T. Yong, "Dark Touristic Perception: Motivation, Experience and Benefits Interpreted from the Visit to Seismic Memorial Sites in Sichuan Province," in *Journal of Mountain Science*, 11, no. 5 (2014): 1326–41; A. Brian et al., "Consuming Post Disaster Destinations: The Case of Sichuan, China," in *Annals of Tourism Research* 47 (2014): 1–17.
30 "Nepal Maoist Leader Prachandra Opens 'Guerrilla Trail,'" BBC, October 2, 2012, accessed March 22, 2016, www.bbc.com/news/world-asia-19815779.

31 Y. Mansfeld and T. Korman, "Between War and Peace: Conflict Heritage Tourism along Three Israeli Border Areas," *Tourism Geographies*, 17, no. 3 (2015): 437–60.
32 S. Robins, "Constructing Meaning from Disappearance: Local Memorialisation of the Missing in Nepal," *International Journal of Conflict and Violence* 8, no. 1 (2014): 104–18.
33 J. J. Lennon, "Tragedy and Heritage: The Case of Cambodia," *Tourism Recreation Research*, 34, no. 1 (2009): 35–43.
34 B. A. Wielde Heidelberg, "Managing Ghosts: Exploring Local Government Involvement in Dark Tourism," *Journal of Heritage Tourism* 10, no. 1 (2015): 74.

12 From living memory to social history

Commemoration and interpretation of a contemporary dark event

Elspeth Frew

Memorialisation of disasters

When a tragic event occurs, care needs to be taken to ensure any memorialisation associated with that event is designed appropriately, which acknowledges that members of the public have lost close friends, family or colleagues in the tragedy and are grieving and coming to terms with the atrocity. As the years pass, and as the tragic event gradually moves from contemporary memory to social history, the site should be designed to be able to make this transition, and ensure it has the potential to retain its relevance for future generations. A contemporary dark event is one which has occurred in living memory, such as the Lockerbie air disaster in 1988 or the 9/11 terrorist attacks in the United States in 2001. The contemporary nature of these 'hot events' can make interpretation at these sites difficult, as the events cannot easily be placed in a larger historical continuity and context.[1] This chapter explores the extent to which the on-site commemoration of the Lockerbie air disaster has the potential to move from contemporary commemoration to a site of significant social history, particularly now that nearly thirty years have passed since the incident and the twenty-fifth anniversary has been commemorated.

When an emergency occurs and people have been killed there are a range of immediate aspects that need to be dealt with, such as the recovery of the remains and the identification of the victims.[2] However, following the emergency, the family and friends of the victims need to be able to grieve, and there should be recognition that traumatic stress can be suffered by many members of the community who have been exposed to the disaster, such as in the 9/11 attacks, the Bali bombings and the terrorist attack in Oslo and on Utøya Island, Norway. Following such tragic events, discussion occurs about the most appropriate public memorials to develop for the site, as happened for example following the 2005 London bombings, the Columbine massacre and the earthquake in Christchurch.[3]

Grief in society has been examined to determine the best means to support people who are experiencing loss and the most appropriate way to memorialise tragic events.[4] Local authorities have to consider the best ways to allow the public to grieve and this can be a challenge for authorities and governments who need to respond appropriately to perceived community needs.[5] For example, following the terror attacks of July 2011, Norwegian health authorities piloted a new model whereby each survivor was given individual tailored help, a local contact person and reunions for the survivors and their families.[6]

Anniversaries of tragic events are particularly important in the grieving process and grief counsellors note the importance of acknowledging the anniversary of the loss of someone, particularly because recognising and acknowledging feelings that may surface around the anniversary of the death of a loved one can be a crucial part of the recovery process.[7] Such rituals are important as they may help individuals deal with their grief by providing structure and stability, and can add meaning to the experience of loss.[8] Bereaved individuals are often encouraged by health professionals to recognise anniversaries and to find a way to acknowledge their significance.[9] Thus, survivors and families of lost loved ones may appreciate the efforts made by institutions, communities and government authorities to stage a commemorative event in recognition of the significance of the anniversary and the part it can play in the healing process.[10] Additionally, such rituals create a sense of community and give bereaved individuals an opportunity to receive acknowledgement, support and acceptance from others.[11]

Following a disaster there is often evidence of a range of post-disaster activities among the grieving public to remember those killed as a way to express personal and collective emotions, such as shock, anger, disbelief and grief.[12] In recent times public outpouring of emotions have been recognised which have been captured in such outlets as condolence books, social media such as Facebook and Twitter, spontaneous shrines, temporary memorials, roadside memorials or grassroots memorials.[13] Local authorities should be aware of this outpouring of emotion and find ways for the visitors to express themselves and allow them to spend quiet, contemplative time at the site of the tragedy, particularly as people feel they have the right to remember the dead in a variety of ways.[14] This private grief in public spaces reflects aspects of bereavement theory and as such needs to be handled sensitively.[15] Once the initial grieving period has passed, the discussion turns to the role of the memorial sites in the future, as their purpose changes from being a place to express grief and facilitate mourning to a place of reflection and remembrance.[16] There is also recognition that travelling to the place where the tragedy occurred can help in the healing process and allow people to come to terms with their loss and to provide some healing via appropriate memorialisations.[17] The emergence of dark tourism as a social practice whereby people visit places of past mass deaths has been facilitated partly due to 'contemporary economic, technological and communicative capabilities that stretch over pre-existing national boundaries and international relations'.[18] The desire to visit such sites needs to be taken into account by authorities to allow citizens from a variety of nationalities to visit the sites of such tragedies.

The Lockerbie air disaster

The Lockerbie air disaster occurred when Pan Am flight 103 exploded over the town of Lockerbie in south-west Scotland on 21 December 1988. A total of 259 passengers and crew members as well as eleven of the town's residents lost their lives. It was one of the worst crashes in aviation history and resulted in possibly the highest number of deaths by terrorist act until the 2001 attack on the World Trade Center.[19] The debris and human remains were scattered over a large distance; there was an extensive search for survivors and then, when it was established that there were no survivors, for human remains. Most of the passengers were US citizens returning to the US for the Christmas holidays, and students from Syracuse University were among the victims, with the students all aged about twenty. The two Libyan citizens accused of blowing up Pan Am 103 were extradited to the Netherlands on 5 April 1999 to stand trial.[20] On 31 January 2001, a panel of three Scottish judges found Abdelbaset Al-Megrahi guilty of planning and executing the bombing and sentenced him to life with a minimum of twenty years in prison. He was released in 2009 as he was terminally ill with prostate cancer, and died in 2012.[21] The Lockerbie disaster transformed British attitudes to aviation security.[22] Since the bomb was placed in a piece of luggage that was not x-rayed or reconciled with its passenger, the following changes were made to aviation security: 100 per cent of baggage is now screened at all major British airports, reconciliation of passengers with their cabin baggage is now in place and security of cargo and supplies has been addressed.[23]

The local community of Lockerbie has staged commemorative services each year since the air disaster, but in particular commemorations have been held on the significant dates of the first, tenth, twentieth and twenty-fifth anniversaries to honour and remember those who died.[24] On the first anniversary in 1989, the relatives of the crash victims laid wreaths in front of the memorial in the Garden of Remembrance, which had been built at the Dryfesdale Cemetery, Lockerbie and was dedicated by local clergy in a simple ceremony. Later that day a second service was held and included a traditional lament played by a Scottish bagpiper outside Lockerbie's town hall, which had been used as a temporary mortuary. At the Tundergarth Church, just outside the town, families of victims attended a service at a small chapel near where the airliner's cockpit had landed.[25] At the tenth anniversary there was a church service at the Dryfesdale Parish Church, attended by more than 700 people. This service included a minute's silence at 7.03pm, which was the moment when the bomb exploded on the aircraft. There were simultaneous services at Westminster Abbey in London, Arlington National Cemetery near Washington and Syracuse University in New York State. Relatives of sixty victims travelled to Lockerbie and 'were joined by residents who have become friends'.[26] Messages were read out from the Queen, the UK Prime Minister Tony Blair and US President Bill Clinton. The Duke of Edinburgh, local councillors, friends and relatives laid wreaths at the memorial at the Dryfesdale Cemetery. Father Pat Keegans was reporting as saying at the cemetery:

> Some would tell us that we should draw a line now at your tenth anniversary. We shall not draw any line. To do so would be an insult to your lives, to your families and friends and to the people of Lockerbie. You will see us laying wreaths at your stone. We want you to be sure that these wreaths are not hollow, empty gestures but a statement and declaration full of promise. It is a

declaration that we will not rest until we have justice and truth, until all responsible for your deaths are held accountable.[27]

This statement indicates that in 1998 no one had been found guilty of the crime. At the twentieth anniversary, more than 150 people gathered for a wreath-laying ceremony at the Dryfesdale Cemetery. There was also a service in the chapel at London's Heathrow Airport, from where Pan Am Flight 103 took off. Again, a minute's silence was held in Lockerbie and at Heathrow at 7.03pm to mark the exact time of the tragedy. As with previous anniversaries, in the United States ceremonies were organised in Arlington National Cemetery in Virginia and at Syracuse University in New York State.[28] Similarly, on the twenty-fifth anniversary, simultaneous remembrance services took place at Lockerbie, Westminster Abbey in London, Syracuse University and Arlington National Cemetery in the US. Relatives read the names of the victims following a minute's silence at 7.03pm (2.03pm EST in the US). British Prime Minister David Cameron was reported as saying:

> Though 25 years have passed, memories of the 243 passengers, 16 crew and 11 Lockerbie residents who lost their lives on that terrible night have not dimmed. Today our thoughts turn to its victims and to those whose lives have been touched and changed by what happened at Lockerbie that night.[29]

Graham Herbert, former rector at Lockerbie Academy, which lost three students in the atrocity, continued: 'I know today there will be a lot of closed doors. A lot of people will not go out of their houses. The memories are just too bitter, there are still open wounds there.' Jane Schultz lost her twenty-year-old son Thomas, who was part of the Syracuse University group on board the flight. She said:

> In my heart, to me this is home and there was no other place I felt I should be on this very sad and special occasion. I wanted to be here to honour my son as well as the 269 other victims and I wanted to stand in the place where my son took his last breath and say a small prayer.

The simultaneous services provide a symbolic means of tying the various communities together in their remembrance of the victims and help symbolise and reinforce the bonds between the families of the victims created following the events of 21 December 1988.

As well as the anniversary commemorations, the town of Lockerbie and its surrounds have developed the following memorials to the air disaster: a Garden of Remembrance and a Visitors' Centre at the Dryfesdale Cemetery; a memorial at the Tundergarth Church which contains a book of remembrance and a visitors' book; memorials at Sherwood Crescent and Rosebank Crescent, where local lives were lost and houses destroyed; and a memorial stained-glass window at the Lockerbie Town Hall. This chapter gathers detail from the Visitors' Centre and the Garden of Remembrance, and assesses evidence of contemporary memorialisation and the potential for the site to move forward into social history now that nearly thirty years has passed since the disaster occurred.

Garden of Remembrance

Just outside the town of Lockerbie is the local cemetery, dating back to the 1730s, which contains the Dryfesdale Lodge Visitors' Centre and Garden of Remembrance. The Visitors' Centre is located in a traditional Scottish lodge, built using local sandstone, and was renovated in 2003 to accommodate visitors to the cemetery, including the addition of ramped access and public toilet facilities. The lodge is open from March until September and also one week prior to the anniversary on 21 December. Near the Garden of Remembrance a plaque states: 'This Garden of Remembrance is in memory of the 270 victims whose ages ranged from 2 months to 82 years from 21 nations.' The Garden of Remembrance is set away from the traditional gravestones and consists of a triptych grey granite plaque which lists the names of all the victims. The memorial stone is built into the traditional stone wall of the cemetery and is surrounded by individual memorial plates. To each side of the memorial there are park benches, and planted flowers form two semi-circles in front of the memorial. There is a pathway around the flowers which leads the visitor to the memorial.

The memorial states: 'In remembrance of all victims of Lockerbie Air Disaster who died on 21st December 1988.' The stone is divided into three sections and the list of all the victims covers six columns. The names are listed in alphabetical order by surname so it is easy to identify victims with the same surname from possibly the same family. For example, a father, mother and child with the same surname provides a poignant reminder that some of the victims were family members travelling together. As well as the main memorial there are numerous individual plaques provided by the family, friends or colleagues of those who died. The individual plaques are different styles and sizes and contain a range of sentiments. For example, some plaques contain no dates, while others have the date of birth but no date of death. The plaques are evenly spaced along the traditional cemetery walls and in the grass around the memorial, but their variety of sizes, styles and content reflects the diversity and number of victims. Some of the plaques contains a photo or sketch of the person who died. For example, there is the engraved image of a young man named Turhan Michael Ergin, who was twenty-two years old when he died. The plaque contains an engraved image of his face, below which is a quote attributed to him when he was fourteen which states: 'Never forget how to laugh, because laughter is the inner energy which cleanses the soul and lifts the spirit.' Similarly, another plaque shows a sketch of Suzanne Marie Miazga, also aged twenty-two when she died; beneath her name and age it states 'I want to live life, not just exist'. These two examples illustrate the type of sentiment of the plaques and reinforce the age and innocence of the victims. The inclusion of such sketches adds a human face to tragedy and helps visitors relate to the loss of human life (see Figure 12.1).

Figure 12.1 Memorial Plaque and Visitor Centre, Lockerbie, Scotland.
Photo copyright E. Frew.

The Visitors' Centre

The Visitors' Centre consists of three rooms. The local history room contains information about the early history of the area, the Romans, the Dark and Middle Ages, local families and the growth of Lockerbie, and includes two panels on the Lockerbie air disaster. The second room is available for local people to display arts and crafts. The third room is the Remembrance room, which contains a wide range of artefacts, newspaper cuttings and memorabilia to remember the victims of the tragedy. The two panels in the local history room provide a matter-of-fact reflection on the events of the day. The first one is headed 'The Lockerbie Air Disaster 1988' and explains in plain, non-sensational language the sequence of the events, describing the explosion, the impact of the explosion on the ground and the responses from the emergency services and the local people. The following paragraph makes strong reference to the human elements of the event:

> The enormous scale of the disaster produced a remarkable response. Regional and District Council officials, the police, the fire service and medical services went into action immediately. Voluntary help in many forms from both near and far provided all kinds of assistance ... A dreadful and distressing task was the recovery of the dead.

The panel describing the air disaster includes a series of seven photos taken on the day or soon after the tragedy which show destroyed houses, a silhouette of emergency services workers and bare trees in front of a house completely ablaze, the nose-cone of the downed airliner in a field with rescue workers in the background and an aerial shot of a large crater among destroyed suburban houses. The second panel is headed 'The Lockerbie Air Disaster – The Aftermath' and it states:

> The very moving Memorial Service on 4th January 1989 was attended by many dignitaries including the Prime Minister, Mrs Margaret Thatcher, and the Leader of the Opposition, Mr Neil Kinnock. At a later date HM The Queen and HRH the Duke of Edinburgh visited the town and in particular the memorials in Dryfesdale Cemetery.

Under the photos an extra piece of paper has been added to the panel with sellotape, reflecting its recent addition, which states: 'Abdelbaset Al Megrahi was freed from Greenock prison on compassionate grounds in August 2009. He had prostate cancer and was expected to live for only three months. In fact, he survived for well over two years, finally dying on 20 May 2012.'

Thus, in the Visitors' Centre there is evidence of attempts to engage the visitor directly with content that has personal meaning via the use of strong human-interest themes to encourage reflection.[30] In addition, the use of photography helps the display to be more powerful and to depict the enormity and seriousness of the event.[31] The mention of the visits and messages by dignitaries shows that when an event has national and international significance, high-profile dignitaries often visit the site to offer comfort and solace.[32]

The role of creativity as therapy for people suffering grief has been found to have a positive role in helping people to deal with their loss.[33] Reflecting the importance of creativity and the arts in the memorialisation and grieving process, a poem is provided in the Visitors' Centre, written by Doreen Moscrop. The poem runs for sixty-one lines and is entitled 'Lockerbie'. The following nine-line extract from the poem highlights the tragic event and the involvement of the local people:

> *The fires and the carnage*
> *Were a sickening, fearsome sight.*
> *For some it was the shortest day*
> *December the twenty-first*
> *For some it was the longest day. . .*
> *The only glimmer of light that shone*
> *From that little Scottish town*
> *Was the courage of its people*
> *As they walked around*

The therapeutic properties of quilting have been found to be helpful for a variety of reasons, as have the benefits of poetry and creativity in general.[34] The Remembrance room is dominated by a large quilt which is attached to the wall opposite the entrance to the room. Under the quilt a notice states:

> This quilt is dedicated to all those who lost their lives in the events of 21 December 1988. The 259 leaves on the tree depict the souls who lost their

Figure 12.2 Quilt in Remembrance room and tree in Arts and Crafts room, Visitor Centre, Lockerbie, Scotland.

Photo copyright E. Frew.

lives whilst passengers on Pan Am Flight 103. The 11 pebbles depict the residents of Lockerbie who lost their lives (see Figure 12.2).

Contemporary memorialisation

The arts and crafts room of the Visitors' Centre is dominated by a cardboard tree attached to the wall which features many paper leaves. A note beside the tree explains that the local primary school students studied the events of 21 December 1988 and were asked to write about their feelings after visiting the cemetery. Each of the 259 leaves on the tree has a phrase written on it by a primary school student in response to three different statements, namely 'How I felt when I visited the Garden of Remembrance', 'What I noticed at the Remembrance Garden' and 'Why it is important to remember the Lockerbie Disaster'. The following quotes are representative of the types of sentiments written by the local primary school children:

How I felt when I visited the Garden of Remembrance

– *Really sad*
– *I felt sorry for everyone who died*

What I noticed at the Remembrance Garden

– *There was flowers down to remember all the people who lost their lives*
– *I noticed that there was lots of names on the very, very, very, big grave stone*
– *Lots of flowers and names*
– *People were planting flowers*

Why it is important to remember the Lockerbie Disaster

- *Because my Dad has told me about them. My Dad's friend was called Steven and he lost his Mummy, Daddy and sister Joanne*
- *Because a lot of people died*
- *To remember the people that died*
- *Because it was in Lockerbie that so many people died – it's what Lockerbie is known for now*

Many of these local school children are the children and grandchildren of Lockerbie residents who may have experienced the event first-hand, and the residents may still have painful memories. This reinforces the importance of these children learning about the tragedy too. This creative method of recording the thoughts and feelings of the local children ensures the exhibit is contemporary and significant, and reinforces the current and ongoing poignancy of the site.

Another example of the site being personal and contemporary is experienced via a memory board in the Visitor's Centre which is covered in different coloured Post-it notes written by visitors. This is an informal way of acknowledging individual visitors' thoughts and feelings and supports the research on memorialisation and condolence books.[35] At the time of visiting, a notice on the memory board stated: 'If you would like to leave a message for the 25th Anniversary, please feel free to do so'. The messages on the twenty-fifth anniversary Post-it notes were written by either close family members of the victims, neighbours of the victims, friends of the parents of victims or members of the general public who had no personal connections with the site but had visited the Visitors' Centre and felt compelled to write a message. The following message written on a Post-it note suggests that the mother and sister of the deceased still suffer deeply from the loss of their son and brother:

> John Ahern O'Connor – 25 years without you has been – you know – still think of you. Miss you every day. Thanks for brightening our lives while you were here. Mom cries for you all the time – we all do – love you so much. Can't wait to join you. Love Bonnie.

Similarly, the following message from a police officer connected with the tragedy suggests that the event remains relevant to him:

> Stationed at East Kilbride police office on 21/12/88. Worked at Lockerbie Ice Rink in the aftermath of the crash. I was profoundly affected by the events at Lockerbie and this is my first proper visit since then. These people are always in my thoughts and will always be remembered. RIP.

Thus, the types of sentiments expressed on the Post-it notes in 2013, during the year of the twenty-fifth anniversary, demonstrate that people continue to feel strong emotions associated with the event, despite the passage of time.

In the small hallway connecting the three rooms in the Visitors' Centre there is a wall dedicated to information about the relationship between the Lockerbie residents and Syracuse University, New York State, as thirty-five students died in the disaster when they were returning from studies in Europe. The information in the hallway states: 'Every year two pupils from Lockerbie Academy spend a year attending the University of Syracuse, New York State. The link between the university and the community of Lockerbie continues as a bond of mutual friendship.' A page taken from the web site of the 'Office of Undergraduate Studies Lockerbie Scholars' at Syracuse University displays the names of two students each year who have travelled to the university from 1990 to the present day, representing more than fifty young people over the past twenty-six years (at the time of writing). This reflects the established and ongoing links with the university though these Lockerbie families and the university community. The exhibition at the Visitors' Centre noted that the families of victims established the following five organisations after the tragic event: 'Families of Pan Am 103/Lockerbie', 'Justice for Pan Am 103', 'Lockerbie Friendship Group', 'UK Families Flight 103' and 'Victims of Pan Am Flight 103, Inc'. Support groups have been shown to be important to help people in the grieving process[35] and the establishment of these organisations would have allowed the grieving relatives to focus their energy on engaging in positive activities, which has continued for more than twenty-five years post-disaster.[36]

The memorialisation activities in the Garden of Remembrance, in the Visitors' Centre and during key anniversary events described above have similarities with other sites commemorating death and disaster around the world. For example, memorials have been built and commemorative events staged on significant dates, such as the first, fifth and tenth anniversaries, at Ground Zero, commemorating the 2001 9/11 attacks in the US, and at Port Arthur in Australia, in remembrance of the massacre in 1996. At such commemorative events there is often a minute's silence at the exact time when the event occurred, after which the names of the victims are read out. The anniversary may also feature the laying of wreaths in front of the memorial; a service may be held, with scripture readings and hymns sung, and high-profile dignitaries may attend. Such commemorative events may involve symbolic activities such as releasing doves or ringing a bell to mark the day. Thus, there are many similarities between the Lockerbie Garden of Remembrance and Visitors' Centre and other disaster sites, reflecting the universality of some aspects of memorialisation.

Concluding comments

When an event has occurred in living memory, initially families and the general public are engaged with the event, as they are grieving and the tragedy is fresh in the minds of the public. However, as time moves on and the event becomes family memory, and then community memory, consideration should be given as to what needs to be done to ensure the memorial remains relevant. The aim of this chapter is to explore the extent to which the on-site commemoration of the Lockerbie air disaster has the potential to move from contemporary commemoration to a site of significant social history. The chapter has considered the ways in which the Lockerbie Garden of Remembrance and the Visitors' Centre have engaged in contemporary memorialisation over the course of twenty-six years and the extent to which it has the potential to move into social history in the future.

This chapter has demonstrated that the Lockerbie Garden of Remembrance and Visitors' Centre use imagery, text and language in the commemoration of the disaster. The site mirrors certain aspects of memorialisation common at other sites around the world, such as providing respectful, formal language in the interpretative panels, but also includes some informal language via the Post-it notes and the inclusion of children's voices via their writings. The use of the children's voices helps reinforce that those who died were the innocent victims of a terrorist attack and allows a younger generation of visitors, such as school children, to express their sorrow and sadness regarding the incident.

A close examination of the displays in the Visitors' Centre also reveals several themes in the interpretation and commemoration of the Lockerbie disaster. For example, the use of original photos taken moments after the debris hit the village creates authentic and dramatic images, helping to illustrate the size and intensity of the event. Photos of dignitaries visiting the site reinforce the seriousness and significance of the occurrence, and strengthen the ways in which many people (particularly in the UK and US) were grief-stricken and sought solace. At the Visitors' Centre the elements which ensure the memorial remains relevant are created via the personal reflections of the families of the victims, the local people who lost neighbours and friends and the personal reflections of the emergency workers who faced terrible conditions when they arrived to help. The focus on individuals' involvement in the search-and-recovery process recognises the impact that the event had on the local population. The use of creativity through artistic pieces such as the quilt suggests a therapeutic aspect of the site for visitors, with the symbolic inclusion of trees to represent life and lives lost in the fallen leaves. The Remembrance room itself is very peaceful, allowing the visitors to move around at their own pace and think about the significance of the tragedy, and to remember those who died. The non-sensationalist aspect of the Visitors' Centre and the quietness of the cemetery provide the conditions to allow visitors to experience quiet reflection and contemplation.

In the Visitors' Centre there is also recognition of the establishment and continuation of the strong link between the people of Lockerbie and Syracuse University via the Lockerbie Scholars over the past years. The Visitors' Centre displays recent articles and news items about the tragedy in the Remembrance room and includes mention of books written or articles published about people who died. In addition, the next generation of local children, who have no first-hand memory of the tragedy but are aware of it from families and schooling, wrote about their own experiences and feelings, and the display helps visitors reflect on the continued strength of emotion in the town. The Visitors' Centre allows recognition and acknowledgement for the relatives and friends of the victims, who continue to attend the officially sanctioned annual memorial events held in the local church and simultaneously at other locations. As such, the Garden of Remembrance and the Visitors' Centre appear to have used techniques to continue to ensure the memorial has relevance for the locals and visitors, despite the passage of time.

Notes

1 D. L. Uzzell, "The Hot Interpretation of War and Conflict," in *Heritage Interpretation, Vol. 1: The Natural and Built Environment*, ed. D. L. Uzzell (London: Belhaven, 1989), 35.
2 A. Enander, "Human Needs and Behaviour in the Event of Emergencies and Social Crises," in *Emergency Response Management in Today's Complex Society*, ed. L. Fredholm and A. Göransson, 2010, accessed March 8, 2016, www.msb.se/RibData/Filer/pdf/25880.pdf.
3 C. Chung and Y. M. Easthope, "Traumatic Stress and Death Anxiety among Community Residents Exposed to an Aircraft Crash," *Death Studies* 24, no. 8 (2000): 689–704; J. Hamblen and L. B. Slone, *What Are the Traumatic Stress Effects of Terrorism?* (Virginia: National Center for Posttraumatic Stress Disorder, Department of Veteran Affairs, 2001); J. Hawdon and J. Ryan, "Social Relations that Generate and Sustain Solidarity after a Mass Tragedy," *Social Forces* 89, no. 4 (2011): 1363–84.
4 C. Corr and D. Corr, *Death and Dying, Life and Living* (London: Cengage Learning, 2012); H. Dauncey and C. Tinker, "Media, Memory and Nostalgia in Contemporary France: Between Commemoration, Memorialisation, Reflection and Restoration," *Modern and Contemporary France* 23, no. 2 (2015): 135–45; K. J. Doka, "Memorialization, Ritual and Public Tragedy," in M. Lattanza-Licht and K. J. Doka, eds, *Living with Grief: Coping with Public Tragedy* (New York: Brunner-Routledge, 2003), 179–90.
5 N. P. Kropf and B. L. Jones, "When Public Tragedies Happen: Community Practice Approaches in Grief, Loss, and Recovery," *Journal of Community Practice* 22, no. 3 (2014): 281–98; S. Nicholls, "Disaster Memorials as Government Communication," *Australian Journal of Emergency Management* 21, no. 4 (2006): 36–43.
6 Kärki, F. U. (2015). "Norway's 2011 Terror Attacks: Alleviating National Trauma With a Large-Scale Proactive Intervention Model," *Psychiatric Services* 66, no. 9 (2015): 910–12.
7 K. Jordan, "What We Learned from the 9/11 First Anniversary," *The Family Journal* 11, no. 2 (2003); D. G. Nemeth et al., "Addressing Anniversary Reactions of Trauma through Group Process: the Hurricane Katrina Anniversary Wellness Workshops," *International Journal of Group Psychotherapy* 62, no. 1 (2012): 129–42.
8 Kropf and Jones, "When Public Tragedies Happen."
9 E. Kübler-Ross and D. Kessler, *On Grief and Grieving: Finding the Meaning of Grief through the Five Stages of Loss* (New York: Simon and Schuster, 2005).
10 A. Eyre, "In Remembrance: Post-disaster Rituals and Symbols," *Australian Journal of Emergency Management* 14, no. 3 (1999): 23–9.
11 P. Dass-Brailsford, "Disasters," in *Social Death. The A–Z of Death and Dying: Social, Medical, and Cultural Aspects*, ed. M. Brennan (Santa Barbara: ABC-CLIO, 2014), 161–5.
12 Eyre, "In Remembrance."
13 J. Rosenberg and D. L. Peck, "Individual Reactions and Social Responses to Massive Loss of Life," in Clifton D. Bryant, ed., *Handbook of Death and Dying* (Thousand Oaks, CA: SAGE, 2003), 223–36; J. Santino, ed., *Spontaneous Shrines and the Public Memorialization of Death* (Basingstoke: Palgrave Macmillan, 2006).
14 I. Kelman and R. Dodds, "Developing a Code of Ethics for Disaster Tourism," *International Journal of Mass Emergencies and Disasters* 27, no. 3 (2009): 272–96; D. Viejo-Rose, "Memorial Functions: Intent, Impact and the Right to Remember," in *Memory Studies* 4, no. 4 (2011): 465–80.

15. K. Woodthorpe, "Private Grief in Public Spaces: Interpreting Memorialisation in the Contemporary Cemetery," in Hockey et al., ed., *The Matter of Death: Space, Place and Materiality* (Houndmills: Palgrave Macmillan, 2010), 29–132; K. Woodthorpe, "Using Bereavement Theory to Understand Memorialising Behaviour," *Bereavement Care* 30, no. 2 (2011): 29–32.
16. A. King, *Memorials of the Great War in Britain: The Symbolism and Politics of Remembrance* (London: Bloomsbury Publishing, 2014); J. M. Winter, *Sites of Memory, Sites of Mourning: The Great War in European Cultural History* (Cambridge: CUP, 1995).
17. N. Drvenkar, M. Banožić, and D. Živić, "Development of Memorial Tourism as a New Concept—Possibilities and Restrictions," *Tourism and Hospitality Management* 21, no. 1 (2015): 63–77.
18. R. McManus, *Death in a Global Age* (Chicago: Palgrave Macmillan, 2012), 171.
19. K. I. Matar and R. W. Thabit, *Lockerbie and Libya: A Study in International Relations* (North Carolina: McFarland, 2003).
20. Ibid.
21. D. Kenealy, "Commercial Interests and Calculated Compassion: The Diplomacy and Paradiplomacy of Releasing the Lockerbie Bomber," *Diplomacy and Statecraft* 23, no. 3 (2012): 555–75.
22. P. Wilkinson, *The Lessons of Lockerbie: A Special Report on Aviation Security to Mark the First Anniversary of the Air Disaster* (Report no. 226, Research Institute for the Study of Conflict and Terrorism, 1989).
23. O. Malik, "Aviation Security before and after Lockerbie," *Terrorism and Political Violence* 10, no. 3 (1998), doi: 10.1080/09546559808427473.
24. K. H. Rodoski, "An Enduring Legacy of Hope: Twenty-Five Years after the Pan Am 103 Terrorist Bombing, SU and Lockerbie, Scotland, Remain Committed to Honoring Those Lost and Moving Forward in Their Memory," *Syracuse University Magazine* 30, no. 3 (2013): 10; D. Britton, "Elegies of Darkness: Commemorations of the Bombing of Pan Am 103" (PhD diss., Syracuse University, 2008).
25. G. Elgood, "One Year On, Families Remember PanAm Disaster victims," *Reuters News*, December 21, 1989.
26. G. Harris, "Ten Years On, A Nation Remembers," *The Times*, December 22, 1998, 10.
27. Ibid.
28. AAP Bulletins, "Ceremonies Mark Lockerbie Anniversary," *Australian Associated Press Pty LT,* 2008.
29. Sky News, "Special Memorial Services to Mark the 25th Anniversary of the Lockerbie Bombing have Taken Place in England, Scotland and the US," *Sky News*, December 21, 2013.
30. G. Black, *The Engaging Museum: Developing Museums for Visitor Involvement* (London: Psychology Press, 2005).
31. I. Dekel, "Ways of Looking: Observation and Transformation at the Holocaust Memorial, Berlin," in *Memory Studies* 2, no. 1 (2009): 71–86.
32. A. Eyre, "Remembering: Community Commemoration after Disaster," in Rodriguez et al., ed., *Handbook of Disaster Research* (New York: Springer, 2007), 441–55.
33. S. L. Bertman, ed., *Grief and the Healing Arts: Creativity as Therapy* (New York: Baywood, Amityville, 1999); R. A. Neimeyer, *Techniques of Grief Therapy: Creative Practices for Counseling the Bereaved* (London: Routledge, 2012); J. E. Rogers, ed., *The Art of Grief: The Use of Expressive Arts in a Grief Support Group* (London: Routledge, 2011).
34. D. Britton, "Comfort in Cloth: The Syracuse University Remembrance Quilt," *Voices* 34, no. 3/4 (2008): 3; V. A. Dickie, "Experiencing Therapy through Doing: Making Quilts", *OTJR: Occupation, Participation and Health* 31, no. 4 (2011): 209–15; W. R. Dunton Jr, "Quilt Making as a Socializing Measure," *American Journal of Physical Medicine and Rehabilitation* 16, no. 4 (1937): 275–8; C. Malone, "The Art of

Remembrance: The Arts and Crafts Movement and the Commemoration of the British War Dead, 1916–1920," *Contemporary British History* 26, no. 1 (2012): 1–23.
35 P. J. Margry and C. Sánchez-Carretero, "Rethinking Memorialization: The Concept of Grassroots Memorials," in *Grassroots Memorials: The Politics of Memorializing Traumatic Death*, ed. P. J. Margry and C. Sánchez-Carretero (Oxford, Berghahn Books, 2011), 1–48; M. Brennan, "Condolence Books: Language and Meaning in the Mourning for Hillsborough and Diana," *Death Studies* 32, no. 4 (2008): 326–51.
36 G. J. Stevens et al., "Coping Support Factors among Australians Affected by Terrorism: 2002 Bali Bombing Survivors Speak," *The Medical Journal of Australia* 199, no. 11 (2013): 772–5.

13 Experiencing dark heritage live

Britta Timm Knudsen

I visited the island of Utøya in Tyrifjorden, Norway, on 27 September 2014, on one of the first occasions that the island was open to the general public after the mass shooting that took place on 22 July 2011.[1] Disguised as a policeman, extreme right-wing perpetrator Anders Bering Breivik went on a shooting spree on the small island hosting a yearly summer camp for the youth organization of the Norwegian Labour Party, killing sixty-nine people and wounding another sixty-six. I believe my own urge to go to this crime scene was in order to experience something of the event for myself, and to try to understand what motivates many tourists to travel to dark sites. In a sense, dark tourism is always about connecting with ghosts, paradoxically embodying what is no longer there, but making it physically present and tangible through affects: the feel of more or less remote, and certainly more or less worked through, pasts. The dark tourist is also an interesting, paradoxical figure in a modernist Western paradigm of all-pervasive visibility, surveillance and theatricality.[2] She/he is part of the developing creative industries, as well as the tourism and heritage sectors, and, whether the dark tourist wants it or not, she/he is part of an industry that thrives on spectacle and staging. On the other hand, the dark tourist is looking for something that is not there, for the possibility of relating to the ghosts of the past. Although there are all sorts of motivations to be dark tourists, I believe the primary motivation is to connect with something that is no longer there.

Dark tourism is not an entirely new field: from Seaton's 'thanatourism'[3] and Tunbridge and Ashworth's 'dissonant heritage',[4] through Lennon and Foley's generic text, to Sharpley and Stone's multifaceted edited volume of 2009, there have been many contributions to the field.[5] With these developments thematic diversity has been studied, with war tourism, genocide and slavery tourism all discussed, as well as the incorporation of a wider set of theoretical underpinnings. From Marxist perspectives, which often criticize consumerist (mass) tendencies in dark tourism,[6] to the poststructuralist performative turn which establishes the ability to comment on, co-produce and eventually alter the sites' significations – either through innovative site-design or actual tourists' alternative practices – many interpretive strategies have been used.[7] More recently, dark tourism has embraced a more materialist, post-human stance, putting relations and affective relationality between a 'now' and a 'then' on the agenda, as a replacement for purely constructivist paradigms.[8]

This chapter explores the more-than-representational and emotional aspects of dark tourism, especially through the concept of 'liveness', here interpreted as a feeling of being in the immediacy of events, even though they belong to the past. The chapter will focus on the ways in which liveness is produced in the experience of contemporary dark tourists who inevitably visit dark sites *after* the historically significant event has occurred, and will explore various forms of liveness production *before* coming to the sites, *at* the sites and in the tourists' own co-production of the sites *after* their visit. In this way I subscribe to the idea that dark tourism has become a global culture industry with a particular cultural logic of immersion between media and things, image and matter.[9] This entails a dynamic relationship between places and mediated representations that are both extensive and intensive; sites are disseminated extensively – globally – through media, and they are experienced intensively – corporeally – due to their characters as immersive media environments.[10] As a visitor to the 'mother' site of terrorist attacks, the World Trade Center in New York, I experienced at first hand a dark tourist site that took the consequence of its own status as *lieu de mémoire*, and the large number of media witnesses that came, and still come, to the site, seeing in reality what they had experienced through the media.'[11] By 2004, Ground Zero was already a fully fledged experiential design taking place around the tomb site of the missing towers, with survivors, widows of rescuing firefighters and local residents acting as guides, inviting the dark tourists into the event as equal and global media witnesses of it. My point is that dark tourism lives, paradoxically, off liveness, and this chapter will analyse the many forms of liveness appearing in the cases on which I have worked for the past ten years.

Liveness

It may seem contradictory to focus on the concept of liveness in dark tourism, as we have just stated that dark tourists are historically invested modern travellers. But, as I will argue, this is not the case. Performance theorist Philip Auslander[12] defines classic liveness as a series of significant characteristics: 'Physical co-presence of performers and audience; temporal simultaneity of production and reception; experience in the moment.'[13] From this point of view, dark tourism is a post-event activity that has missed the chance to be present in both the time and space of a dark event; is referred to as the 'dead'; and is represented in museums, memorial sites, monuments and shrines. However, classic liveness has lost its original meaning of equivalence to physical presence through developments within broadcast technologies throughout the twentieth century, and has become especially compromised by terms such as 'live broadcast' and 'recorded live'. In the first case, performers and audiences are temporally but not spatially co-present, and in the second, performers and audiences share neither a temporal framework nor a physical location.[14] The two latter forms of liveness appear paradoxical, but they are now widely accepted terms that show how the definition of liveness has changed significantly with the emergence of new communication technologies. Auslander states in a constructivist manner that 'live is actually an effect of mediatization, not the other way around'.[15] 'Mediated' and 'live' are mutually dependent, and almost always co-exist in all sorts of experiences. Thus, the concept of liveness gains at least two new meanings different from that of classic liveness, which we see in the following two statements: 'The liveness of the experience of listening to or watching the recording is primarily affective: live recordings allow the listener a sense of participating in a specific performance'; and 'However, the word "live" has also come to refer to connections and interactions with non-human agents'.[16] Auslander's broadening of the concept of 'live' to refer more to a *feeling* than to an ontological state thus means that if audiences feel as if they are part of something from which they are distant, in either time or space, then the experience *is* live. All sorts of non-human agents (technologies, relics, atmospheres, etc.) play the role of mediators and producers of liveness in actual dark tourism encounters.

This line of thought continues and underscores the broadening of concepts similar to liveness in actual media scholarship, such as that of witnessing, a category we may use to describe the function of dark tourists.[17] In Ricœur's *Mediawitnessing: Testimony in the Age of Mass Communication*, media witnessing succeeds his phenomenological concept of 'eyewitnessing', meaning being there, experiencing a dark event not only through the eyes but through the whole body, sometimes at risk.[18] This is to say that although dark tourists can never become eyewitnesses to past events, they may get a *feeling of having been there* through modes and media at dark sites.[19] For a dark tourist who is absent in time but present in space, only historical representations seem available. Through the renewed concept of liveness we may add that the dark tourist has to have the feeling of participating in the absent dark event, and has to interact with non-human actors relating to the event in order to be touched. Using Bolter and Grusin's concept of 'remediation' – oscillating between immediacy and hypermediacy, between the experience of the real and the experience of the medium – Astrid Erll connects liveness and mediated-ness to memory culture in general, and states that immediacy – liveness – is achieved through representational build-up.[20]

Liveness before getting to a site

Media representations of dark sites are important factors in reconfiguring symbolic landscapes within which tourism occurs. Media co-construct dark events – such as wars, conflict, terrorist attacks – as they unfold in actual media representations, and distribute symbolic positions. They play a central role in people's desire for and actual purchase of trips to certain destinations, but there is more to it. Besides the general agenda-setting power of media – news, literature and global movie products – which Couldry has characterized by highlighting tourism to film sets (set-jetting) as media pilgrimages, something else is at work.[21]

In order to understand the liveness factor of these media representations and why they are able to engage people, one has to consider their moving capacities. One way of proceeding is to turn to what Linda Williams calls 'body genre movies'.[22] Here, 'body genre' means a genre that has an actual physical impact, meaning that the affective impact may be quite clear: melodramas that make you cry; horror films that give you goosebumps, cause sudden physical movements or make you cry out; or pornography which arouses you sexually. We can transfer body genres to media representations and say that representations have excited and electrified our bodies, have induced in us a potentially strong desire to relive these emotions. Actual travel to dark sites – for example, a detour to Josef Fritzl's house in Amstetten, Austria – could be capable of triggering the horror that one feels at the events played out in that house. This is not to claim that social reality is lived along the lines of fiction, or that people are not capable of distinguishing between horror movies and real horror; on the contrary, there is in fact proof of a very (self-) aware and strategic manipulation of one's own psycho-social state. Such manipulations of one's own affects, which Grossberg saw in teenage viewing communities of genre movies, are at work in tourism.[23] One goes to dark sites in order to experience a certain mood, to (re)connect with the seriousness, tragedy and injustice of life conditions.

My research trips to Auschwitz-Birkenau, Ground Zero[24] and Sarajevo,[25] as well as my work with online documentation of the unfolding violent events in the Ukraine,[26] share the common features of being at dark sites in which liveness factors are generated through their iconic status of having been media images in a number of representations, in various media. And as such they appear as live places, in various ways and through different media. For example, the extermination camp of Auschwitz-Birkenau has been especially perpetuated through photography and documentaries, whereas the siege of Sarajevo was a much-televised event on news features from 1992 to 1995. Ground Zero was the first terrorist event to be documented live through amateur and semi-amateur videos and mobile devices, a tendency that only escalated in later terrorist events or natural disasters (the Indian Ocean tsunami of 2004, the Madrid train bombings of 2004, the London bombings of 2005, the attack on *Charlie Hebdo* in Paris in 2015).

Media can create 'new compulsions of proximity' that reconfigure the landscape within which tourism occurs. But, more importantly, remediations of already recognized dark tourism sites are able to vitalize them and provide them with new significations, or awaken an interest in forgotten, denied and repressed dark pasts. Despite their fictional statuses, movies and television series may create new interest in historical events, and vitalize tourism to certain dark tourist sites. The television series *1864* (released in 2014), on the Danish–German border war, increased the number of tourists to the entrenchments in Southern Denmark, and the recent movie *The Gold Coast*, on Danish colonialism in Western Africa, most certainly will encourage dark tourists to go to Ghana to follow in the footsteps of Danish slave trade history. Thus, visual media – regardless of the content's fictitious character – are able to vitalize places, and encourage tourists to travel. Tourists also travel to sets and key places associated with crime fiction, and again, this is not to say that dark tourists do not distinguish between fact and fiction in their tourist behaviour; it is more the case that remediations encouraging dark tourism actualize the media–materiality entanglement that characterizes intensive global culture, and thus produces new liveness for acknowledged dark heritage sites as well as for new sites. The feeling of liveness and the ghost-haunting is the same.

Liveness at dark sites

The primary medium through which one experiences a dark tourism site is the place itself. The *in situ* character of the place itself is the most important liveness factor of all. Huge, global historic events of an atrocious nature – such as the slave trade, the First and Second World Wars and communist terrorist regimes in the Soviet Union or Kampuchea – have their places and scarred landscapes. For example, the First World War trenches in Ieper, which are perfectly restored remains in the natural landscape, enable the contemporary dark tourist to experience the soldiers fighting there and, as such, to cultivate the former's sympathetic imagination of the latter's 'living' conditions.[27] Sympathy is exactly what we feel when we engage with figures from the past.[28] The historical dark 'situations' may be more or less elaborate, more or less staged. I will continue by pointing out three places of dark tourism that evoke sympathy, by using different degrees of liveness.

Let us recapitulate liveness within the context of dark tourism. Liveness happens when the contemporary tourist lives the 'same' situation as the historic figure, and when the dark tourist can make her/his sympathetic imagination work to feel the situation of the other, or in this case the historic other. Dark tourism sites have to put the contemporary tourist in relation to the historic other in order to be able to evoke sympathy and change awareness, as well as possibly views and behaviours. This may happen in multiple ways, through various media: the medium of an *in situ* place itself, for example, visiting industrial heritage sites such as coal mines in Dortmund, a 3D representation of the results of the German bombing raids during the Warsaw Uprising in 1944 at the Warsaw Uprising Museum or architecture and artistic representations working at the materiality of places, such as the Memorial for the Murdered Jews in Europe in Berlin and Jewish Museums in Copenhagen and Berlin. The use of live witnesses or descendants of live witnesses at dark sites (prisons, death camps, slave plantations) is a frequent and efficient liveness producer. Here, the narrative, voice and body of the living witness are the media that place the contemporary tourist within history. But likewise, the media that produce liveness may be more staged and overtly immersive, such as in re-enactments or stagings of historic situations in particular circumstances.[29] Let us look at two examples.

A theatrical production took place between December 2009 and March 2010 in a two-storey relic from the Cold War, a bunker situated twenty-five kilometres outside Vilnius. The re-staging was entitled *Deportation Day: Live History Lesson*, and involved a staged re-enactment in various sequences: the transport to and reception at the camp; the interrogation; distribution of equipment; labour in the form of snow-shovelling; the meal. The play was directed by Lithuanian theatre critic Ruta Vanagaité; it featured two Russian actors and a dog, and all the props used – in the form of posters, audioscapes, clothing, utensils, furniture and so on – were original relics. Various groups took part: European students from exchange programmes, Lithuanian students, ordinary tourists and school children. This immersive environment was exclusively created in order to evoke sympathy in contemporary visitors, as we can see from the following quotation from Ruta Vanagaité:

> Anyone can be broken. This is a very educational point for young Lithuanians and foreigners alike. You are no different from Lithuanians. If you find yourself in a place from which there is no escape, and if you are shouted at in a language you don't understand, you quickly start to feel that you are nobody – which is exactly the feeling we had.[30]

The play obviously tries to establish sympathy with the subjected and imprisoned Lithuanians through a re-enactment of an extraordinary situation of labour-camp internment. However, what is actually hoped for through the imitation of the internment and de-subjectifying of the interned is a kind of kinaesthetic awareness that breaks the self–other divide, in order to catch glimpses of sameness.[31] A similar experience may be said to exist for dark tourists at historically significant sites: a desire to break the self–other divide in order to feel the complicated heritage of the other. In this difficult heritage case, the contemporary tourist is therefore managed according to the desire of the producers.

In the second example of cultural heritage tourism in the post-communist countries we have a much looser management, coming from the fact that dark tourism design stems from local entrepreneurs without any formal and institutionalized educational obligations. Crazy Guides Communism Tours to Nowa Huta and the steelworks – Stalin's gift to Krakow, and now home to 250,000 people – are run by a group of young Poles managing the legacy of former generations with the explicit aim of presenting an alternative to mainstream tourism to Krakow.[32] The aim to be alternative is driven by the need to earn money based upon this marketable heritage, but not exclusively, since the motivation is to also establish sympathy with the communist other through a series of staged-real situations in Nowa Huta. Such tours are quite comprehensive: they include rides in polluting and noisy Trabants and visits to four iconic sites in Nowa Huta: the central square Plac Centralny – renamed the Ronald Reagan Square in 1989; the entrance of the huge Tadeusz Sendzimir Steelworks; a Russian tank available as a solitary relic in public space; and the Church of Our Lady, built by locals as a protest, and tolerated by the Russians despite the ban on religion in communist times. Two immersive environments appear in the programme. One is a visit to the Stylowa Restaurant next to Rose Avenue, which appears as a time warp, and during which tourists hear the guides' narrative on everyday life in Nowa Huta from the 1950s onwards. The second is a mimetic environment of an apartment from the 1950s–70s in Nowa Huta. In this instance visitors are served gurki and vodka in a fully furnished three-room flat, with the opportunity to watch a propaganda movie about the wonders of Nowa Huta while listening to the guides' criticism of the movie's narrative.

Crazy Guides Communism Tourism, as a dark tourism site, is much more complex and non-linear in its communication with the 'dark tourists' than the labour-camp site is. This may be attributed to a range of factors, including *experiential meaning potential*, the surplus of value produced in the situation itself.[33] I would suggest three circumstances that produce liveness in this case. First are the crazy guides themselves, who are young, fun and entrepreneurial, and who are important assets in this tourism design. Not only do they become live witnesses of a dark legacy, they also become interesting in their own right: what is the future of these young Poles? Their ironic and self-ironic tone (aimed toward stereotype-chasing tourists and their consumption habits at the same time that they are also supporting those) is crucial, as it plays out as a more open, political identity agenda in post-community societies that are neither communist nor capitalist, and may be more precisely labelled as 'becoming' – another strong liveness marker. Second, the fact that their tourism design plays itself out in a live, everyday environment, where ordinary and extraordinary Poles live their lives, is an important liveness marker. The third circumstance producing liveness is that the staged-real character, especially of the apartment, actualizes the tourists' various memories of their own childhoods and thus escapes the Cold War distinction between East and West, and an exotic representation of the communist other.

Liveness after visiting a dark site

Liveness experienced at a dark site and liveness produced after a visit to a dark site are connected by the huge liveness marker around places in the process of becoming dark tourism sites that is created by the witnesses' own documentation and recording of dark events as they unfold. Since 9/11, witnesses' own amateurish recordings on their mobile media devices have been a most valuable source of mobilization, as their immediacy is unrivalled.[34] They are extensively used in electronic mass media, shared through digital media as forms of witnessing that are able to document events in cases of trials and denying authorities' narratives and also capable of mobilizing broader audiences.[35] Live witnessing not only plays a role in sites coming into being as future dark tourism sites, but tourists' post-productions of established dark tourists' sites are important in order to revitalize the desire for those sites.

The documentation of visits carried out by dark tourists is important in tourism in general. These forms of post-productions of sites cover a range of visual productions, from photos and selfies to YouTube videos uploaded and shared on various social media. The transformation from a media user paradigm to a media producer paradigm that is developed in concepts of participatory and DIY culture is likewise at play in dark tourism.[36] Here, liveness before going to a site and liveness after having visited a site are connected, as tourists' post-productions of dark sites are dependent on the scale of media coverage during dark events. My study of YouTube videos dedicated to Sniper Alley, as the street Ulica Zmaja od Bosne was informally baptized during the forty-four-month-long siege of Sarajevo in 1992–6, is such a case.[37] To stand on the street in contemporary Sarajevo has the liveness quality of seeing for oneself a haunted place. Many tourists physically revisit the places they have experienced through their media experiences, make their own productions and share them on YouTube.

Such tourist productions may be divided into three categories. First, we have the uploading of found footage that re-mediates a presence in both space and time.[38] These videos serve as relics of the event, and get by far the most views and viewer comments, as they are unrivalled in connecting audiences with a particular mood. Second are videos that are new productions, present in space, but absent in time. These are also capable of stirring debate, such as the video made by American tourist Glenn Campbell, who uploaded a short video showing him pointing to Sniper Alley and to the buildings on the left side of the road, claiming:

> This road I am biking on was Sniper Alley during the war, the most dangerous place in Sarajevo. The radical Serb forces held those apartments over there [the camera tilts to the left and shows huge building blocks] with snipers in the windows shooting at everything that moved. You had to use this road because it connected to the airport and down town. So people drove there at an enormous speed because the Serbs were shooting at everybody.[39]

Campbell received much hate mail in relation to that twenty-seven-second video, and the comments showed that political discussions continue to flare in this unfinished conflict.

Third, we have videos that represent a high degree of production, often produced by semi-professionals in the form of video testimonies of their encounter with the city: the higher the degree of elaboration, the fewer views, and thus the less interest and liveness factor. The first category of video establishes an affective mood of being there; the second category rehearses the political positions and emotions involved in the conflict, which are still open-ended; and the last category shows Sarajevo and Sniper Alley as a more general, poetic topic in tourists' productions. The liveness factor of tourists' post-production of dark tourist sites, therefore, plays out in two ways. On the one hand, post-production itself is an expression of the emotional impact that the site has had on the individual tourist; on the other, liveness is also the immediate emotional impact, and its ability to stir up heated debates that these post-productions have on contemporary audiences in a kind of open-ended, democratic deliberation.[40]

My last case in this travel through liveness stays within the YouTube arena, and focuses on war memorials in the form of video tributes to fallen Danish soldiers in recent conflict zones as an online commemorative practice.[41] Dark tourism is of course not only something that implies physical travel to sites, but also something that plays out as an online activity. Terrorists' videos of beheadings and war-zone testimonies may be considered online dark tourist sites. And most certainly, online war memorials are dark tourism sites, as they function as online visual and textual expansions of their offline equivalents, such as war tombs or war memorials. The video tributes are to be understood as two-fold vernacular responses to national events such as any grassroots memorialization, whether it takes the spontaneous form of laying down flowers, poems and lights at death sites or takes the form of more developed online productions.[42] They are two-fold because they consist of both an audio-visual part – in the form of a short video with an average length

182 *Britta Timm Knudsen*

of four minutes, produced by fellow soldiers, relatives or friends showing the deceased in all kinds of situations, quite often reinforcing dominant official discourses and perspectives on the war – and comments that often present a quite deconstructive and critical force vis-à-vis the videos. The videos represent the memorial, and the comments represent the dark tourists' rendering of their experiences of these memorials 'en direct'. The two-fold memorial thus has the outstanding possibility of being a living, incomplete memorial in the making, presenting the reactions to the tributes in the form of political dissent from the official stance on warfare as a more general comment that is not directed at the deceased, individual soldier. One could argue that these kinds of two-fold memorials are more democratic than their offline counterparts, but what is more important to the argument here is the fact that they guarantee liveness as an integrated part of the online vernacular memorials.

The liveness of live situations

Dark tourism can also just play itself out in media, as dark events to a large extent are documented, shared and disseminated online. My last example will focus on a forty-seven-second long audio recording of the Copenhagen shooting in 14 February 2015. This piece of audio recording has been largely disseminated and thus it is worth looking at its liveness-producing features. The attack on the French weekly satirical newspaper *Charlie Hebdo* in Paris on 7 January 2015, during which the gunmen brothers killed eleven persons present at the offices and injured another eleven, was widely documented by amateur videos of the events in their immediate unfolding. The same was the case with the Copenhagen shooting event on 14–15 February 2015 during which the perpetrator, a twenty-two-year-old Dane of Palestinian descent, killed two persons and wounded several policemen at a freedom of speech event and in front of a synagogue in the city later the same day.

Because of their close relation in time, the two events became linked among the Danish public. French Ambassador François Zimeray was present at the free speech debate in the Danish capital, as well as Swedish cartoonist Lars Vilks, who has drawn caricatures of the Prophet Muhammad. But the two countries are also conceptually linked. Freedom of speech – at the expense of minority protection – has been on the agenda in both countries since the Danish cartoon crisis in 2005. In February 2006, *Charlie Hebdo* decided to print the twelve Muhammad cartoons in solidarity with the publication of the twelve drawings in Danish newspaper *Jyllands Postens* on 30 September 2005. The freedom of speech solidarity between the two countries was re-affirmed through the 'Je suis Charlie' identity-political reaction expressed in many solidarity commemorations expressed in the aftermath of the event in Denmark, mirrored bby the front page of French newspaper *Liberation* on Monday 16 February 2015 reading 'Vi er Danskere' (translated as 'Nous sommes Danois', 'We are Danes'). This was an intertextual reference to the 9/11 terrorist attack on the WTC that gave birth to the expression that 'We are all Americans', which again was a citation of the phrase 'We are all German Jews' from the Second World War. This rhetorical feature shows an overall solidarity with the victims and expresses the shared victimhood and vulnerability of all human beings.[43]

These mirroring actions in the two countries show a similarity in the discursive understandings and framings of the violent events: it is all about freedom of speech and all about not respecting freedom of speech. I would like to try to answer the question as to why this particular piece of audio recording, from the free speech event at Krudttønden in the eastern district of town, has travelled and what distinguishes it from the many amateur immediate video recordings of the events both in Paris and in Copenhagen. Many of the video recordings show quite violent images of wounded, dying or dead corpses during the events, taken from a safe position, normally above the events at street level. The audio recording lasts forty-seven seconds. We hear Femen activist Inna Shevchenko talking about freedom of speech:

> I realize that every time we talk about the activity of those people, it will be always: Yes, it is about freedom of speech but. And the turning point is but. Why do we still say but, when we ... [Interruptions of gunshots, around thirty in all; chairs rattle; sounds of iron bars hitting the ground; a low voice commanding: down; distant yells from the street].

The differences between the video recordings and this audio recording regard the distance from the event, the perspective on the event and the recording body's implication in the scene. The live recording takes place in the middle of the event; no images emerge from it – expressing the lack of overview that people experiencing the event must have had – and the recording body is an endangered body, heightening its legitimacy as a credible witness. These features give the audio recording a very trustworthy impression of how it must have felt to be there, increasing the liveness factor. On top of this, the situation almost plays out the original scenario of speech being interrupted by violence that prevents dialogue and speech taking place, thus facilitating the freedom of speech framing of the event as a whole.

Conclusion

I now wish to return to a phrase used at the beginning of this chapter, when I suggested that dark tourists are modern travellers in search of ghosts from the past and referred to *the feel* of the past as an important factor in dark tourists' experiences. Sharpley and Stone ended their book by speculating about the motivations of dark tourists: 'are people drawn to [the exhibition of] Body Worlds by a ghoulish fascination', they wrote, 'or by a search for meaning?'[44] To this question I would suggest that the non-representational layers of a site – online or offline – are absolutely crucial, and in line with Hans Ulrich Gumbrecht's notion of spatial presence and the production of presence as something that alternates the search for 'meaning', I would argue that liveness in the forms presented here is important in order to feel the event and the violent death conveyed.[45]

Instead of investigating further the motivations of dark tourists, the time has come to look at the impact of dark tourist practices and desires at concrete sites, in relation to concrete debates and on established power geometries. Some of the cases presented in this text show that new transnational alliances are sometimes made through the mediated character of sites, either online or offline. What dark tourism evokes affectively in tourists, however, is the very paradox of bodily safeness and common vulnerability and the unequal distribution of these positions historically and in contemporary societies. Experiencing that paradox live could be a step in a learning-to-be-affected endeavour that urges mobilization and action around important issues.

Notes

1. B. Timm Knudsen and J. Ifversen, "Commemoration, Heritage and Affective Ecology – The Case of Utøya," in *Affecting Heritage*, ed. D. P. Tolia-Kelly, E. Waterton, and S. Watson (Aldershot: Ashgate, 2016), forthcoming.
2. These ways of interpreting modernity or late modernity as a visibility regime are mutually exclusive and highlight various aspects of this epoch: M. Foucault's *Surveiller et punir* (Paris: Gallimard, 1975) and P. Virilio's *La Machine de Vision* (Paris: Galilée, 1988) define 'subjecthood' as 'being visible and thus potentially being disciplined or erased by the gaze of power'.
3. A. V. Seaton, "Guided by the Dark: From Thanatopsis to Thanatourism," in *International Journal of Heritage Studies* 2, no. 4 (1996): 234–44.
4. J. J. Lennon and M. Foley, *Dark Tourism: The Attraction of Death and Disaster* (London: Continuum, 2000).
5. R. Sharpley and P. Stone, eds., *The Darker Side of Travel. The Theory and Practice of Dark Tourism* (Bristol: Channel View, 2009).
6. T. Cole, *Selling the Holocaust: From Auschwitz to Schindler; How History is Bought, Packaged and Sold* (London: Routledge, 2000); M. Sturken, *Tourists of History: Memory, Kitsch, and Consumerism from Oklahoma City to Ground Zero* (Durham: Duke University Press, 2007).
7. A. Jackson and J. Kidd, eds., *Performing Heritage: Research, Practice and Innovation in Museum Theatre and Live Interpretation* (Manchester: Manchester University Press, 2011); S. Magelsen and R. Justice-Malloy, eds., *Enacting History* (Tuscaloosa: The University of Alabama Press, 2011); M. Daugbjerg, R. Syd Eisner, and B. Timm Knudsen, eds., "Re-enacting the Past," in *International Journal of Heritage Studies* 20, no. 7–8 (2014): 742–59.
8. J. Sather-Wagstaff, *Heritage that Hurts: Tourists in the Memoryscapes of September 11* (Walnut Creek, CA: Left Coast Press, 2011); E. Waterton and S. Watson, "Methods in Motion: Affecting Heritage Research," in *Affective Methodologies*, ed. B. Timm Knudsen and C. Stage (London: Palgrave, 2015), explore two dark tourist sites: the Towton Battlefield in York, with the highest rate of deaths in British history, and the Uluru Kata-Tjuta National Park in Australia, which is a World Heritage Site for the indigenous other within a nation-state having a colonial past.
9. S. Lash and C. Lury, *Global Culture Industry* (Cambridge: Polity, 2007).
10. Ibid., 9.
11. Media scholarship has broadened this concept, which stems from P. Nora, *Les Lieux de Mémoire* (1984–92) to also cover immaterial media representations of dark events. Audiences become media-witnesses, meaning that they, as well as locally placed eye-witnesses, witness the event through media: P. Frost and A. Pinchevski, eds., *Media Witnessing. Testimony in the Age of Mass Communication* (London: Palgrave, 2011).

12 P. Auslander, *Liveness Performance in a Mediatized Culture* (London: Routledge, 1999, 2008).
13 Ibid., 61.
14 Ibid., 60.
15 Ibid., 56.
16 Ibid., 60–1.
17 See note 11 above.
18 P. Ricœur, Lectures 3, *Aux Frontières de la Philosophie* (Paris: Seuil, 1992).
19 It is quite possible to become a dark tourist to current wars and ongoing conflicts, often expressed through the adventure travel concept – transporting tourists into ongoing war activities such as those unfolding in the Ukraine, or through audiences' more political interests. See the following websites, accessed 8 March 2016: www.warzonetours.com; www.mickendsor.com/companies/war-tourism-industry; www.politicaltours.com
20 A. Erll, *Memory in Culture* (London: Palgrave, 2011).
21 N. Couldry, "On the Actual Street," in *The Media and the Tourist Imagination*, ed. D. Crouch, R. Jackson, and F. Thompson (London: Routledge, 2005), 60–5.
22 L. Williams, "Film Bodies: Gender, Genre, and Excess," *Film Quarterly* 44, no. 4 (1991): 2–13.
23 L. Grossberg, *Dancing in Spite of Myself: Essays on Popular Culture* (Durham, NC: Duke University Press, 1997).
24 B. Timm Knudsen, "Thanatourism: Witnessing Difficult Pasts," *Tourist Studies* 11, no. 1 (2011): 55–72.
25 B. Timm Knudsen, "The Besieged City in the Heart of Europe: Sniper Alley in Sarajevo as Memorial Site on YouTube," in *Mediating and Remediating Death*, ed. D. Refslund Christensen and K. Sandvik (Aldershot: Ashgate, 2014), 111–33.
26 B. Timm Knudsen and C. Stage, "Media Witnessing from Chora," in *Global Media, Biopolitics and Affect. Politicizing Bodily Vulnerability*, ed. B. Timm Knudsen and C. Stage (London: Routledge, 2015), 131–53.
27 In his book on the Holocaust experience, Gary Weissman distinguishes between pity and sympathy, saying that 'pity enables us to feel *for* the survivor, the "sympathetic imagination" enables us to feel *like* the survivor': G. Weissman, *Fantasies of Witnessing: Postwar Efforts to Experience the Holocaust* (Ithaca: Cornell University Press, 2004), 110.
28 Using a comparison with Adam Smith and his theory on sentiments, we can say that the distinction between pity and sympathy is explained further. Whereas pity and compassion signify our fellow-feeling with the sorrow of others, sympathy denotes our fellow-feeling with any passion whatever; and, more importantly: 'Sympathy, therefore, does not arise so much from the view of the passion, as from that of the situation which excites it.' A. Smith, *The Theory of Moral Sentiments* (London: Penguin, 2009 [1790]), 16.
29 Reenactments as live media are used both in citizen-driven and entrepreneurial heritage activities (Gettysburg, WWII, etc) and in artistic versions of dark events, such as Company Ea Sola's re-enactments of everyday experiences and memories of the Vietnamese–American war. See the special issue, "Re-enacting the Past," of *International Journal of Heritage Studies* 20, no. 7–8 (2014) and the documentary *The Act of Killing* (2012), by Danish–American director Joshua Oppenheimer, in which former Indonesian death-squad leaders are encouraged to re-enact their former deeds for contemporary audiences to watch.
30 This interview appeared in B. T. Knudsen, "Deportation Day: Live History Lesson," in *Museum International* 63, no. 249–250 (2011): 109–18.
31 Kinaesthetic awareness is a concept taken from Nigel Thrift's *Non-Representational Theory: Space, Politics, Affect* (London: Routledge, 2008), 237.
32 www.crazyguides.com (accessed March 8, 2016). See B. T. Knudsen, "The Past as Staged-real Environment: Communism Revisited in The Crazy Guides Communism Tours, Krakow, Poland," *Journal of Tourism and Cultural Change* 8, no. 3 (2010): 139–53.

33 Term from G. Kress and T. V. Leeuwen, *Multimodal Discourse: The Modes and Media of Contemporary Communication* (London: Hodder Arnold, 2001), 22.
34 Witnessing by means of mobile devices has received scholarly treatment by A. Reading, "Mobile Witnessing: Ethics and the Camera Phone in the 'War on Terror'," *Globalizations* 6, no. 1 (2009): 61–76; A. Reading, "The London Bombings: Mobile Witnessing, Mortal Bodies and Globital Time," *Memory Studies* 4, no. 3 (2011): 298–311; K. Andén-Papadopoulos, "Citizen Camera-witnessing: Embodied Political Dissent in the Age of 'Mediated Mass Self-communication'," *New Media & Society* 16, no. 5 (2013): 753–69; A. McCosker, "De-framing Disaster: Affective Encounters with Raw and Autonomous Media," *Continuum: Journal of Media and Cultural Studies* 27 (2013): 382–98; Timm Knudsen and Stage, "Media Witnessing from Chora."
35 H. Jenkins, *Convergence Culture* (New York: New York University Press, 2006); J. Burgess and J. Green, *YouTube* (Cambridge: Polity Press, 2009); Michele Knobel and Colin Lankshear, "DIY Media: A Contextual Background and Some Contemporary Themes," in *DIY Media. Creating, Sharing and Learning with New Technologies*, ed. M. Knobel and C. Lankshear (New York: Peter Lang, 2010), 1–27.
36 H. Jenkins, *Convergence Culture* (New York: New York University Press, 2006); J. Burgess and J. Green, *YouTube* (Cambridge: Polity Press, 2009); M. Knobel and C. Lankshear, "DIY Media: A contextual background and some contemporary themes" in Knobel and Lankshear, eds., *DIY Media*; D. Gauntlett, *Making is Connecting: The Social Meaning of Creativity, from DIY and Knitting to YouTube and Web 2.0* (Cambridge: Polity Press, 2011).
37 B. Timm Knudsen, "The Besieged City."
38 www.youtube.com/watch?v=OeDaObvbE0s (accessed 8 March, 2016).
39 www.youtube.com/watch?v=Qqwb2WPkg2E (accessed 8 March, 2016).
40 An example of a post-production form is re-enactment, a live indication of ongoing interest and passion.
41 B. Timm Knudsen and C. Stage, "Online War Memorials: YouTube as a Democratic Space of Commemoration Exemplified through Video Tributes to Fallen Danish Soldiers," *Memory Studies*, 6, no. 4 (2013): 418–36.
42 P. Jan Magry and C. Sánchez-Carretero, eds., *Grassroots Memorials. The Politics of Memorializing Traumatic Death* (New York: Barghahn Books, 2011).
43 B. Timm Kudsen and C. Stage, *Global Media, Biopolitics and Affect: Politicizing Bodily Vulnerability* (London: Routledge, 2015).
44 R. Sharpley and P. Stone, eds., *The Darker Side of Travel: The Theory and Practice of Dark Tourism* (Bristol: Channel View, 2009), 248.
45 H. Ulrich Gumbrecht, *Production of Presence: What Meaning Cannot Convey* (Palo Alto, CA: Stanford University Press, 2004), xiii.

14 Dark tourism in the brightest of cities
Rio de Janeiro and the favela tour

Glenn Hooper

The marvellous city

Before Brasilia was created as the country's new capital in 1960, Rio de Janeiro had been the focus of Brazil's economic, cultural and political life. Set mid-way between the Brazilian–Bolivian border and the Atlantic ocean, Brasilia was long in the planning, its conception reaching back to the nineteenth century, the idea of a more centrally placed capital away from the eastern corridor an idea that many of the country's politicians had long considered with enthusiasm and pride. In time the city would become famous for its striking and visionary architecture, revered as a landmark of town planning and eventually designated a UNESCO world heritage site, but Brasilia has never really eclipsed Rio.[1] Designed and laid out by an altogether surer hand, Rio is a place at once glamorous and enchanting, compelling in its natural beauty and setting, a city that never fails to impress and delight.[2] Despite the intrusions of colonial and post-independence city planning and development, despite even the mismanagement, political corruption and chronic inequality that has at times sullied its reputation, the city exercises an intoxicating appeal. When the Portuguese Royal Court fled from Napoleonic forces in 1807, they set Portugal and Brazil – as well as Rio – on a very different path. With a retinue of 15,000, the transfer of the Portuguese court to Brazil, where it produced a 'metropolitan reversal' and saw Portugal governed from the New World, created the conditions for Brazilian independence. Rather than continue to regard Brazil as a colony, the monarchy became infatuated with it, and with Rio in particular, where it built palaces and gardens, established schools and hospitals, and in the end simply refused to return 'home'.[3] From a swampy outpost of the Portuguese empire, inhabited by pirates, runaway slaves and criminals, Rio was transformed into a profitable port and city, where coffee and sugar cane were produced, where gold was brought from the interior and sold, and where the granite hills and rainforest were steadily cleared and the shoreline opened.

While it is tempting to think of the history of Rio as one of romantic discovery and settlement, of hardships overcome and improvements made, even of a relatively peaceful transfer of power from Portugal to Brazil, the darker side of Rio is never far away, then or now. Although it is also customary to regard the Brazilians as more racially mixed than their Spanish-speaking neighbours, and therefore more open to the processes of social mobility, hard evidence suggests otherwise. Brazil was in fact the last country in the world to end slavery (in 1888), and the wealth that was held in the hands of some of the city's most powerful families remained inviolate. More particularly, the inequalities that stemmed from those earlier settler developments – of arrival, land confiscation, exploitation of resources and trading – became established patterns, setting Rio on a path of prolonged social but also geo-political disparity. The infamous favelas, the shanty towns that mushroom on the middle and upper reaches of the surrounding mountains, are in many instances still cut off from resources, infrastructure and a genuine connection to the city. The Vidigal favela, for example, sits in the shadow of the Dois Irmaoes Mountain, clearly visible to the surfers and bathers on the world famous Ipanema beach below, yet culturally as well as spatially far removed from that world of luxury, leisure and play.[4] Yet another favela, Rochina – home to anywhere between 100 and 150,000 people – is described as an urban slum, and, although associated with an increasingly confident air of regeneration and potential, is a still challenging space, where drug lords have only recently been deposed under the favela pacification schemes of the last eight years.[5]

While the transformation of Brazilian society over the past thirty years has been the subject of several recent studies, the ramifications of the military dictatorship – materially and militarily supported by the USA, noted for its right-wing ideology, its repressive and authoritarian governance and its widespread torture of dissidents – which held power from 1964 to 1985 have not been entirely dispelled. The move towards free and open elections in the mid-1980s brought about a renewal of Brazilian confidence and hopes for greater economic stability.[6] And indeed, a number of cultural and mega-sporting successes were particularly helpful in re-establishing the city's tarnished reputation, as well as the wider country's profile. Events such as the Copa America football games in 1989 and the Pan-American games which were hosted by Rio in 2007, followed by the announcement that same year that Rio would be one of twelve Brazilian venues which would host the World Cup in 2014, all helped to situate Brazil in a new light. More pertinent to this chapter, perhaps, Rio's tourism industry was also going from strength to strength, partly as a result of a more stable political culture, but also because of dramatic economic growth which, until recent scandals and allegations of impropriety and mismanagement, saw Brazil described as an economic superpower, especially after the discovery of off-shore oil fields. When the city hosts the Olympic Games in 2016 undoubtedly further revenue and accolades will be brought to the city's hospitality and tourism industries, while the wider Brazilian tourism industry, gaining an increasingly surer foothold with its myriad eco-tourism projects, also stands to gain.[7] With a vibrant beach culture and a Carnival held in February of each year seeing a year-on-year rise in the number of overseas visitors, it is hardly an exaggeration to say that Rio would appear to have re-established itself as one of South America's premier destinations: a re-formed city, cosmopolitan in outlook, outward-looking, stable and secure.[8]

When the UN conference on the Environment and Development, with its especial focus on human rights and the alleviation of poverty as outlined in the Rio

Summit, came to the city in 1992 it must truly have seemed that whatever social, political or economic difficulties might periodically materialize, there was simply no going back: that Brazil, and more particularly Rio, were in a very different place. But on the night of 23 July 1993, barely months after the international observers, politicians and journalists had left, members of the Polícia Militar shot dead eight children and wounded several others outside the Candelária Church in the city centre – an event that would have far-reaching consequences in the city, and throw Rio's newly won reputation for reform into doubt. While the city was well used to inequalities and punishments, only too familiar with the heavy-handed tactics of the military and the police, this particular incident, later fully exposed by Julia Rochester in *The Candelária Massacre*, was judged a turning point. As Rochester writes:

> The Church of Our Lady of the Candelária was at the centre of everything, right in the middle of the financial district, within ten minutes' walk from the courts, the museums, the Legislative Assembly and the naval district. It was a landmark just off Rio's busiest shopping street, the Avenida Rio Branco. It was also one of Rio's most important cultural icons, a tourist destination, and it held political significance as the site of pro-democracy demonstrations and trade union rallies. Not least, it was a powerful religious symbol, and the murder of children on a church forecourt was a profound shock to the city.[9]

Not only was this deemed a particularly brazen and offensive outrage, even by Rio's standards, but it also showed that two things remained unchanged: that the police continued to operate outside the rule of law, and that the failures of the favelas, ignored for so long, ravaged by crime, drugs and intimidation, had to be finally tackled.[10] Once again, the city's capacity for violence, as well as its set of long-neglected social and political problems that stemmed from clumsy regeneration initiatives begun in the early twentieth century, were in the news.[11]

In this chapter I wish to discuss the development of a form of tourism that has been criticised and defended in equal measure – which has drawn the attention of academics, politicians and those working in the social services, including education, but which continues to unsettle and rankle. Supported by local and national government, endorsed by local communities, regarded by Brazilian tourist agencies and practitioners as a positive development, the emergence of favela tourism has been seen by many as voyeuristic, distasteful and opportunistic, yet more evidence of a tourist culture pandering to western visitors who have grown tired of beaches and cocktails, cable cars and helicopter rides. Yet the favela tour, while not exactly big business, has flourished in recent years, and there are now ten operators offering trips to various 'pacified' sites.[12] The debate that surrounds the favela tour, then, is both complicated and impacted, and at times as politicised and challenging as the sites themselves. Nevertheless, this chapter will examine the background to the favela tour, discuss its future role within the city's wider tourism offer and consider the favela tour within the parameters of a dark tourism that is undergoing reappraisal even as we write. But we must first begin at the start of the previous century – not because of the need for a historical detour as such, but because an understanding of some of the physical changes that took place in the city helps to better situate the favelas themselves, and, more particularly, shows how so much of what still remains unresolved today has its origins in decisions taken in those earlier decades.

The divided city

Rio de Janeiro has fired the imaginations of many since the start of the twentieth century, bewitching visitors and locals alike. While Carioca elites during the early decades of the twentieth century worked hard at emulating the fashions and tastes of continental Europe through the building of opera houses and wide boulevards, graceful parks and public buildings, there also emerged a sharp division between the privileged and the poor. Exacerbated by an increase in the local population, many of whom had relocated from the interior, already precarious living standards plummeted further when a migrant influx from Portugal and Italy, as well as parts of eastern Europe, was added to the mix.[13] Indeed, Teresa Meade suggests that the population in the final decades of the nineteenth century grew swiftly throughout these years and that in 'Rio proper and the surrounding migrant settlements numbered 518,290 inhabitants at the start of the Republic [1889] and then grew to over one million by 1920'.[14] It is unsurprising that social problems emerged so quickly after such a dramatic increase in population, and indeed comparisons with places much closer to home indicate that Rio was far from being an exception. In the early years of the twentieth century, for example, Dublin was associated with some of the worst tenement slum conditions in the world, where disease, infection and chronic poverty ran rampant, and where the death rate stood, in 1913, at 27.6 per 1000 – a record as bad as Calcutta's. But in Rio there also emerged an additional problem, a physical dissection of the city that would in time produce political as well as social outcomes that would have far-reaching consequences. The city's swamps and mountains may have made the place, as Bruno Carvalho astutely notes, much easier to defend and fortify, but those same 'natural boundaries also constrained its ability to expand – and Rio remains to this day a city in many ways shaped by geographical peculiarities'.[15] Whether we think of the city and, more particularly, its inhabitants in terms of core and periphery, or centre and margin, or privileged and excluded, Rio became increasingly marked by an exceptional social and physical geography that would define its development and identity for years. As Meade suggests of the early twentieth century:

> The city's business and political leaders were appalled at the proliferation of vice, including petty crime, prostitution, vagrancy, gambling, and begging. And without a doubt, the city's topography, wedged between jutting *morros*, the ocean, and the bay, made it difficult to police. But there are indications that the city's elite, then as now, was less concerned with an increase in vice so long as it remained contained to the poorer neighbourhoods.[16]

And, of course, containing those poorer neighbourhoods came to increasingly constitute a whole raft of additional measures: of regulation and isolation, certainly, but also of state security, surveillance and management, much of which further soured already toxic relations between the city's elite and indigent communities. As Jeffrey Needell suggests:

> Such people wanted to put an end to that old Brazil, that 'African' Brazil that threatened their claims to Civilization. And, it was a very present 'Africa' for the elite. Most of the elite had probably been nursed by blacks and were surrounded by black servants, and they had known slavery, abolished only in 1888, at first hand.[17]

So, when the engineers eventually emerged through rock in 1906 and effectively connected the north and south zones of the city together, they did more than bring about something of a technical feat; rather, they directly contributed to the physical reconfiguration of the city, re-orientating the wealth and middle classes away from the old centre, and thereby redistributing the complexity of Rio's social and cultural disparities in new locales. In a very real sense the Túnel Novo connected physical elements of the city together, and broadened the city beyond Botafogo (according to Carvalho, the direction it had been heading in any case). But in another sense they also split it, for as quickly as development and property opportunities were identified in the south zone, the old centre became steadily neglected and run down, its once graceful boulevards and radiating avenues increasingly abandoned for a culture of leisure and fun along what would, in time, become the Avenida Atlântica. Although the development of the south zone was desired by many elites, especially those who felt keenly the physical limits of the city and who hankered for further development opportunities, such shifts were hardly negligible. Carvalho describes the early nineteenth-century efforts that had gone into developing the Cidade Nova in the first place, including new drainage and paving techniques, a public transport system, regular refuse collection and street lighting, as well as the construction of the Mangue Canal (begun in 1857), and 'one of the major infrastructural projects of Imperial Brazil'.[18]

However, the changes that were brought about during the time of Francisco Pereira Passos, engineer and Mayor of Rio de Janeiro in the period 1902–6, transformed the city permanently, and not always for the better. A great admirer of Georges-Eugène Haussmann, the nineteenth-century architect of Paris, Passos' reforms of the city were far-reaching, controversial and irreversible:

> As a result, today we better understand how the demolition of almost 600 central buildings, in what became known as the bota-abaixo, led to a worsening of living conditions for a great number of people and remains deeply interconnected to the growth of favelas. We have gained a clearer sense of the extent to which Brazil's elites equated the notion of civilization to France [. . .] We have also learned how these reforms and the often heavy-handed vaccination campaigns that accompanied them were met with popular uprisings and protests.[19]

There were some undoubted health-related benefits associated with intervention on a scale such as this. Swamps were drained and land reclaimed, smallpox, cholera and yellow fever considerably reduced, and sanitation, beautification and urban renewal did much to transform Rio from a 'city of death' to one of the most attractive and prosperous cities in the Americas, the third major port on the continent, behind only New York and Buenos Aires. But while the physical transformation of the city brought certain benefits, not everyone, not even those who might have been expected to approve, were appreciative. For one thing, a certain level of democratic interaction between different races and creeds, commented favourably upon by many visitors, was lost in the rush for reform. As Teresa Meade suggests, the old city allowed for raucous and relatively unrestrained interaction, for pranks and street dances, all of which were now swept away. 'By the end of the first decade of the twentieth century', she writes, 'almost all of the cabarets and bars where the popular classes occasionally rubbed shoulders with the elite in Lapa and Catete fell under the ax[e] of Mayor Pereira Passos's renovations'.

But much more than the sweeping away of brothels and gaming houses, of disease and rowdiness, the reforms Passos produced would have far-reaching outcomes that would transform the city in ways that not everyone envisaged. In a locale notorious, even then, for its lax morals and almost bacchanalian excess, there developed among certain elements within local government a desire for a culture of respectability and decency, something that could only be achieved if there was greater control and segregation, socially and culturally, but also racially. And here, writes Meade perceptively, 'lay the crux of the issue':

> The concern with civilizing Rio had to do with more than the fear that a few bawdy nightspots or beggars and prostitutes in the main streets of the city discouraged foreign investors and harmed Brazil's prospects as a developing world power. Whatever obstacles these social conditions presented, they paled in comparison with the overriding issue of social control. Entertainment, vice, and social strife overlapped in Rio's downtown, while the potential arose for violent conflict in new and frightening forms.[20]

The public unrest which began on 11 November 1904, when plans to begin widespread vaccination to tackle fever and smallpox brought many onto the streets, demonstrated to the authorities that disaffection, not just disease, was widespread. Indeed, protests were so violent that they had to be contained by cavalry units, as rioters attacked train stations and streetcars, 'cut telephone wires, vandalized and set ablaze streetcars, and broke streetlamps in Catete, Botafogo, and Gávea, zones bordering the center city on the south'.[21] However, while such public unrest was linked firmly to the plans to introduce a universal vaccination programme, wider dissatisfactions had been growing for some time, and resentments over poor sanitation, housing and the rising cost of living, not to mention the unequal development of public services, brought matters to a head.[22] Statistics produced by Meade identify the unequal distribution of city resources and reveal much about the tactless and arrogant plans of a political elite who governed from a position of isolated irresponsibility. For example, the Copacabana district, linked to Botofogo by tunnel, registered no population on the 1906 census, yet was provided, unbelievably and provocatively, with new sanitation and public services. By comparison, the poorer, northern districts of the city saw little improvements made by 1906: 'The thousands of isolated residents of these areas had no sewage system, no lighting, no streets, and no transportation lines to connect them with the rest of Rio.'[23] Such inequalities, underscored by requests from the disadvantaged for better services, improved sanitation, additional resources and housing, would be replayed time and again over successive decades as the favelas multiplied, and their inhabitants' demands and sense of injustice escalated.

The favela tour

Slum tourism, ghetto tourism, township tourism, poverty tourism: all of these are terms which might also be invoked to describe the favela tour. However, I wish to consider it in the context of a somewhat problematised version of dark tourism, because although dark tourism is a term often spurned by national and regional tourist boards for its negative associations, and sometimes by academics for its increasingly elastic and ever fluid parameters, it seems a less demeaning term than many of the alternatives. Also, while Rio's favelas are in some instances slums, with no sanitation or electricity, and where drug lords police neighbourhoods, tuberculosis is still rife and child mortality is shockingly high, some favelas are associated with death, extreme forms of cruelty and, in some instances, torture.[24] More particularly, it would seem that many visiting the favelas of Rio do so as much because of the implicit danger, the lurid stories of gun-fights and reprisals, of death-squads and shoot-to-kill undercover units, and for the reputation many favelas have as 'no-go' areas for police and military both.[25] Coralled in the mountains, sometimes dug into the very rock, clinging to winding and congested roads and along oppressively narrow alleys, here are communities who both exist and don't exist, people who have posed an existential as well as physical threat to the city's political and cultural elite for decades. It is hardly surprising, then, that as tourism opportunities broaden, and tourists themselves rediscover the appeal of the urban, with its issues of poverty, social exclusion and regeneration, that the favela tour, in one of the world's most notoriously divided cities, should emerge.[26]

Opened in 1923, the Copacabana Palace hotel is a five-star venue holding a prime spot on one of the most famous beaches of the world. Faced in white granite, eight storeys high and with a sweeping frontage to allow celebrities to stylishly disembark from chauffeured cars, it evokes art deco glamour and provides high-end luxurious pampering; behind the facade of its elegant architecture, designed by French architect Joseph Gire, may be found gourmet chefs and discreetly attentive staff, beautifully appointed function rooms and lounges, as well as many still intact and original fittings conjured out of exotic hard woods and marble, now blended with carefully chosen art work, ambient lighting and luxury furnishings. In its day it received the most famous and wealthy: Elizabeth Taylor, Marilyn Monroe, Gina Lollobrigida and Brigitte Bardot, to name only a few, stayed at the

Copacabana Palace, where they were fêted and photographed, and where they in turn reflected a glamorous light upon one of the city's most lavish venues. Although it went into a semi-decline in the 1960s and 70s, it has been recently refurbished and is once again regarded as the premier hotel not just in Rio de Janeiro, but in the whole of South America. It was here, at the very entrance to the Copacabana Palace, that I had been told to wait. At 9 am sharp, on an October weekday that was forecast for good visibility and hotter temperatures than were wished for, I was collected by my multi-lingual guide in an unmarked white van to head off into Rio's favelas.

Although the pick-up point was never once mentioned by the tour guide, the emphasis on the contrasting conditions experienced by Rio's poor and the city's more affluent residents was emphasised time and again. Extreme wealth and luxury, architect-designed homes of breathtaking beauty, were pointed out for their glaring juxtaposition with some of the most dangerous and traumatised communities in Brazil. We were a small tour group: a Portuguese couple from Lisbon, an Irish couple from Meath, a single man from Liverpool and myself. As the last of our party was picked up and we threaded our way through the south city traffic, brief introductions were cautiously made among the tourists, though a respectful and vigilant eye was kept towards the front of the van where the guide sat composing himself. What were we there for? My own reasons were pretty clear: to see the favelas, but more importantly to witness a favela tour in action, to assess how it was operated and managed, consider its guiding principles and analyse the behaviour of fellow tourists, as well as to interpret the interactions between tourists and guide and, should we happen to be introduced to any, the locals. Dark tourism is a broad church, as many in this volume have already indicated: from the fairly innocuous traipsings around graveyards, to sites of incarceration, through to the altogether more weighty considerations that are given to battle sites, killing fields and places of extermination. And just to make matters more complicated, the responses evoked by tourists range just as widely: from the scandalised and titillated through to the curious and reflective, before finally arriving at the deeply respectful, where historical and political reality often blends with a sense of shameful disbelief at such scales of human misery.

If dark tourism involves participants deliberately visiting sites associated with 'death and disaster', as Lennon and Foley suggest, or 'death, suffering and the seemingly macabre', as Sharpley and Stone argue, then favela tours qualify on several levels. Life expectancy in the favelas, for example, is much lower than in middle-class neighbourhoods; chronic hygiene, insufficient medical care, poor sanitation and even poorer diets can all take their toll, especially among the very young and the elderly, while mudslides are a constant hazard during the winter months. At least fifty people were killed in mudslides in January 2010, while April of that year saw another four hundred killed and thousands left homeless. However, although death and human suffering is the experience of many favela residents, the lurid stories of running battles among gang members – or, more notoriously, between drug gangs and the police – circulated with mounting hysteria on social as well as mainstream media are possibly an even greater draw for those seeking something out of the ordinary. 'By 1995', writes Misha Glenny, 'the city's murder rate reached a staggering 70.6 per 100,000 inhabitants – not far off Columbia's death rate at the same time, when the Medellin and Cali cartels were at the height of their power'.[27] Seen as tough neighbourhoods where weapons are easily obtained and levels of violence unpredictable and sometimes fatal, for the tourist who has grown tired of conventional tourism offerings, here is a potentially new avenue worth exploring.[28] One might have imagined that after the Candelária Massacre wiser counsel might have prevailed, yet violence only escalated, in some instances to an even greater, more abhorrent extent. Only weeks later, on 29 August 1993, the Vigário Geral favela was to witness further degrees of atrocity, much of it centred on a public bar and involving many innocent victims: 'When the shooting stopped, twenty-one people were dead. The killers were police', writes Julia Rochester almost without surprise: they

> belonged to a death squad calling itself the *Cavalos Corredores* – Running Horses. Over thirty men had participated in the shooting spree. It was a revenge attack, a deeply murky story of the Vigário Geral *tráfico* falling out with police over kickbacks on cocaine shipments.[29]

To what extent does this narrative of violence and counter-violence impact upon today's favela tour? Are the operators and guides simply cashing in on the commodification of favela culture, highlighting its seedier and more perilous aspects in the hope of generating interest and business? And are they, more importantly, responsible for promoting a darker side of the community when more uplifting narratives of comfort and selfless co-operation might just as equally be presented, and just as accurately portray life in the favela? While there is no easy answer to these questions about the need to pander to tourists and to entertain them with heightened stories of violence and danger, the history of the favelas has been one of almost unbroken government indifference, squandered opportunities, chronic inertia and disregard, going right back to those early sanitation and urbanisation schemes. As recently as June 2007, in another favela, Complexo do Alemão, at least forty-four people were killed by the police, with many killings described afterwards as executions. The simple fact, then, is that to tell the story of the favela without explicitly engaging with its violence, irrespective of tourists' expectations, would be a falsehood. From the early twentieth century, when the poor were removed from their houses and deprived of alternative accommodation, when the city's development opportunities drifted south along the coast and when successive governments – military and civilian both – proved unimaginative in their response to the growing, discontented and chronically impoverished slums that ringed the city, violence has been a part of their lives, and its narrative.

The dark tourist

As the minibus climbed upwards, briefly pausing outside the exclusive American school, with its high walls topped with razor wire, our guide explained that security of this level was thought necessary because of the school's proximity to the favela (and we assumed, because of the threat of kidnappings of children of the wealthy). Even though he must have recited all this on countless occasions, his passion for the narrative he told – indeed, seemed compelled to tell – was evident, and he switched seamlessly from a sketch of the recent pacifications schemes, to the percentage of those living without adequate sanitation, to the marijuana and heavy weapons that were routed through the favelas from Paraguay, to the billions allegedly stolen 'by those in power'. Indeed, corruption and inappropriate spending was a common theme, with the World Cup and the upcoming Olympic Games castigated for their exorbitant and spiralling costs, with little evidence, he claimed, that Rio's poorer sections would ever benefit.[30] The guide then broke off, the driver pulled briskly away, and we began to climb steadily up the hill through the exclusive Gávea neighbourhood, with its splendid views of the Christ the Redeemer statue, Corcovado and the city in the distance. 'Prepare yourselves', he announced matter-of-factly, to 'travel from Germany to Ghana in less than three minutes'.

Not unlike the grander dark tourism project, the motivations and ethical issues that surround it, the management of its often highly sensitive sites and the demand-and-supply conundrums regarding which was first to emerge, the favela tour is neither static, easily contained nor convenient to package. Manfred Rolfes, in an analysis of several township and favela tours in Cape Town, Rio and Mumbai, argues that such 'reality' or 'social tours' often have an educational aspect, and that the focus of operators working in this sector is to ensure that the 'town's poor, other or dark side is set at the centre of leisure or touristic activities'. Favela tourism 'is a growing market segment', he goes on to confirm, and, based on research conducted in 2008, he believes the role of many favela guides is to stress 'the relatively high standard of living, the advanced infrastructural equipment, the modern range of services and the varied shopping opportunities' that now exist.[31] In short, he maintains that normality and routine are frequently emphasised because of a perceived need to counter western negativity regarding favela culture and lifestyle. Whether the author had simply had a dramatically different experience or, in the intervening years – which have produced an economic downturn in Brazil – views of favela culture, with its nascent consumerism, have evolved in different ways is hard to say. But my own experiences, with some notable overlap around the subject of poverty, suggests that tour guides are less concerned with glamourising the favela or countering western prejudice than with emphasising the unrest, danger and discontentment that still exists. Yes, some emphasis is laid upon the new facilities recently opened, including some social housing and a bank, but the darker side of life – the hardships, poverty, poor sanitation and ill health of many residents – is a theme that runs throughout.

For a good portion of our tour, indeed, the emphasis is very much on the government's pacification scheme, which was only launched in 2008, the very year of Rolfes' fieldwork. It is perhaps unsurprising that in the years since then there has been some disagreement over the purpose and motivation behind the pacification scheme, with some arguing (logically and convincingly) that its timing was significant; that it followed quickly after the formal announcement in October

2007 that Brazil was to host the 2014 World Cup, and was therefore a political necessity that was simply unavoidable. However, others have argued that high-level discussions around favela pacification pre-dated the announcement, and that it had been in the pipeline for some time, ready for implementation. Whatever the truth of the matter, the latter view, suggestive of long-term planning at senior governmental level, was not one shared by our guide, who was altogether more sceptical. Like the Papal visit of July 2013, sneeringly dismissed as an attempt to entice communicants back from the clutches of Evangelism – an emergent and powerful force in the favelas – the pacification scheme was portrayed as a deliberate, politically motivated obligation rather than a heartfelt attempt at reconciliation and goodwill. And not only that, but the guide repeatedly declared the scheme to have been something of a mixed blessing, because while gang lords had been deposed in some areas – thereby ending a long-standing culture of intimidation – the police pacification unit, the UPP, introduced many new problems, endangered lives and alienated residents, while a by-product of the eradication of drug-related activities was an upsurge in conventional crimes, particularly domestic violence, theft and muggings. 'Who was better?' I asked. 'Neither', came the answer. 'One was as good, or as bad, as the other.'[32]

As has been much discussed throughout this volume, dark tourism is a broad church, with long-established historical precedents, an experiential spectrum of generous and potentially growing proportions and a developing network across very different locations. The dark tourism sites that are the Brazilian favelas have elements in common with those in South Africa, for example, where history and narrative is just as important, and where the dignity of the people is much to the fore. Sixty-nine people were shot dead in 1960 in the Sharpeville Massacre, many in the back as they fled from police, and a Memorial Exhibition Centre now tells the story of that day. But in addition to a specific heritage centre, there are also Sharpeville township tours, just as there are Soweto tours, Apartheid tours and Shantytown tours in Cape Town's Langa and Gugulethu townships. All of these are associated with a history of death and atrocity, and therefore fulfil dark tourism's fundamental criteria as sites of carnage which tourists visit out of curiosity, respect, and so on. However, the dark tourism sites that are the favelas and townships present additional challenges, one of the most demanding being less the ethical issues often cited than the fact that the favelas draw death and life emphatically together. In many conventional dark sites the deaths that are acknowledged are located firmly in the past, with the sites themselves now often museums, memorials and exhibitions. In such places, story boards are constructed, narratives are told and interpretations are offered. In addition, what specifically ties many conventional dark sites together is the fact that death is often presented as a contained historical moment, a narrative legacy that is static and locked, with the viewer rendered a sympathetic but potentially distanced onlooker. The favelas, on the other hand – in being more historically fluid, with sites that are themselves often indistinct and continuous – speak back, are engaged and can even demand more of us in emotional and narrative terms. More importantly, while conventional dark tourism sites may indeed elicit the contemplation of one's own mortality, whether at killing fields or mass graves, the hyper-reality and intensification of authenticity that exists in the favela – ongoing, in process, historically aware yet propulsive – blurs ever more dramatically the lines between life and death. Where many dark sites are fixed, the favela is performative.

We have stopped at the side of road in Rochina favela and have been allowed out, although told to go no further beyond the key-cutter in one direction, and the craft-worker in the other – a fifty-yard stretch at most. Whether this is to allow us to look at the view, to buy from the vendors, to check that we are all comfortable with being outside the cocoon of the minibus, is unclear. We are told that the road, built in the 1950s, was originally laid out as a race-track but is now the main access route through the favela; moto-taxis, motorcyclists who take a pillion passenger anywhere for an agreed fee, whizz by; the now familiar spaghettis of electrical cabling hang dangerously low; dogs lie stretched in the heat; the Portuguese couple strike up what looks like a genuine and amiable conversation with a trader. 'In a favela', writes Teresa Caldeira about São Paulo, although residents sometimes

> build their own dwellings and sometimes pay rent, the residences are constructed on illegally obtained land, and their residents are considered to defy the classification of citizens: they live on usurped terrain, they do not pay city taxes, they do not have an official address.[33]

The anonymity of favela existence described so firmly by Caldeira is one shared earnestly by our guide, who turns constantly to this theme of peripherality and de-classification, drawing our attention to the transient nature of favela life; the endless physical restrictions, the credit that cannot be got due to the debtor having no recognised postal address, and therefore no home that can be easily found.[34] Materially, physically and aesthetically, the guide reminds us, here lies the gross inequality of Rio. There are lavish condominiums, many built in the heyday of the city's 1920s and 30s building boom, with over-wide and generous marble steps, now with doormen, with console tables in the middle distance, sometimes with flowers and curios atop, all nonchalantly ensconced behind reinforced eight-foot-high railings; and favelas, until quite recently constructed of cardboard and plastic, today built of block and cement, yet for all that still derisory and insubstantial. 'Predictably, inhabitants of such spaces are also conceived of as marginal', writes Caldeira perceptively; they are 'considered outsiders: *nordestinos*, newcomers, foreigners, people who are not really from the city'.[35]

While marginality is a topic also developed by Pallmoa Menezes in her study of photographic representations of favela residents, she concludes by viewing the tourist–resident exchange in a much more positive light. Describing her own favela fieldwork as a hybrid experience – a blend of dark and awareness tourism – she sees visitor engagement as affirmative and transformative, much of which she attributes to 'humanisation' and increased interaction between host and guest. Unlike Caldeira, who alleges ongoing prejudice among native Brazilians (middle-class Paulistas and Carioca both), citing their view of favela residents as fixed and irrefutable – an example of geo-political identity conditioned by wider national and historical anxieties – Menezes sees the tourist as capable of change. Locals, it is argued, seem unable or unwilling to regard favelas as anything other than places of dereliction and danger, and persist in that belief; tourists, on the other hand, can move from a position of wariness and unease to one of greater understanding and acceptance. Despite the generally lower expectations many critics have of tourists, the allegations of crass and insensitive cultural arrogance – especially when beyond the cultural confines of western Europe and North America – here at least is an example of affirmative tourist interaction. Of limited long-term political benefit, perhaps, but nevertheless a helpful contribution, Menezes sees the dark tourist as making a positive intervention and having created an instance of meaningful cross-cultural dialogue, whatever literal restrictions are placed upon them by their guides with respect to personal movement and interaction, and despite repeated warnings about the appropriate use of technology.

'Porosity' is the term used by Carvalho to describe the ways in which Rio's urban growth has ebbed and flowed; how the city boundary shifts, the need to accommodate new migrants, the growth of shanty towns and new road layouts, as well as the engineering works that cut through mountainous terrain and extended the city south along the now-famous bays, created a sometimes bad-tempered and uneasy co-existence of wealth and poverty. Of course, many cities are porous to a greater or lesser degree, and one could point to somewhere like Johannesburg, for example – where nervous and gated communities are a similarly widespread phenomenon – as proof that the social and physical liquidity of Rio is not an isolated case. However, the division of the city between *morro e asfalto*, hills and asphalt – between impoverished favelas and bourgeois neighbourhoods – seems somehow more pronounced in Rio, despite the city's capacity to 'absorb elements from the most diverse traditions, across the multiple Afro and Jewish transatlantic diasporas'.[36] Twenty favelas are now 'pacified', our guide informs us, with police operating increasingly like heavily armed social workers, hoping that a broader acceptance of official authority will follow on from this honeymoon period of neighbourliness and improved relations. But how many favelas, pacified or otherwise, still ring the city? We are informed that there are more than six hundred in Rio alone, with 40 per cent of residents without adequate sanitation, where shootouts are still common, and where 80 per cent of those killed during confrontations are desperately poor, male and black.

We travel a short distance, then stop outside a small workshop, and after exiting the van are directed through the shop to the rear of the building, only to find ourselves on a large veranda-type platform. From here we can see much of Rochina tumbling down towards the bays below, but also reaching up into the hills above where, it is said, the police have not fully penetrated and, it is hinted, drug lords still freely operate. The guide is passionate, well informed, articulate and genuine. He produces statistics about unemployment and police fatalities, speaks of ongoing arguments between locals and government officials over healthcare ('we need sanitation, not a cable car!') and discusses briefly the implications of Protestant Evangelical conversion rates. He also appears to handle all questions with a degree of balance and fair-mindedness. For example, the My House/My Life programme, launched by the Brazilian federal government in 2009, and with a sweep of his arm indicated below, is acknowledged as generally well received.[37] At its launch one million low-cost housing units were promised by 2014, and the brightly coloured four-storey apartment blocks with their corrugated roofs that we can see before us are certainly hopeful signs. However, with an economy once more in crisis and attention increasingly focused on the run-up to the Olympic Games in 2016, momentum has stalled, criticism has grown and the true impact of the programme is increasingly seen as negligible.[38]

In an analysis of what he terms 'mortality mediation', Philip Stone draws attention to a number of visitor responses to dark sites that include entertainment, education and memorialisation, moving across the full spectrum of attractions, from light to dark. 'With embodied and emotionally engaged tourists', he argues, 'dark tourism potentially offers the Self an emancipatory place for reassessment and self-reflexivity'.[39] In an experiential approach to dark tourism, Avital Biran and colleagues also argue that 'participants value on-site interpretation as a tool for enhancing their educational experience and knowledge, and as a source for emotional experience and connection to one's personal heritage'.[40] The emphasis on the self developed in both articles is a reminder of the emphatic shift within dark tourism studies that has occurred of late, and of the way in which criticism has increasingly focused more on visitor motivation and perceptions than site management, on agency rather than marketing and other operational matters. Did I believe anyone in our group saw themselves in dramatically different ways, or as having benefited particularly from these experiences? When we stopped later in the morning in the middle of a busy road and were asked to get out, then issued with instructions for walking in single file, like a military foot patrol, through a neighbourhood of food vendors, hardware shops and open sewers, did I sense that a sense of self had become especially pronounced? I have to say that I noticed nothing appeared to have changed much among my companions, save the usual educational and experiential benefit that comes from any organised tours. For myself it was somewhat different, though I suspect different mainly because of my role as a tourism researcher and academic on the lookout for motivations and experiences – primed with a self-conscious if not politically aware sense of what I was partaking in, mindful of sensitivities and trained to look, listen and take note(s).

During the entirety of our tour we were told about mortality and death, made aware of human suffering and gross inequality, provided with statistics concerning disease, poverty and health. In other words, we were at a dark tourism site whether we liked it or not; we were in attendance at what Rojek calls a 'black spot' a significant tourist attraction that is also 'a death site'.[41] In Misha Glenny's analysis of Rochina favela he reminds us about police death-squads and revenge murders, about body parts that are dismembered and then burnt ('No body. No crime'), and he painstakingly describes the violence that has plagued the favelas until relatively recently. 'In terms of overall recorded deaths', he writes, 'this was a low-level conflict. But in the favelas, the homicide rates were comparable to those of countries at war'.[42] That tourists should wish to visit such places undoubtedly raises all sorts of questions, and as we wound our way around Canoas favela, now described by our guide as urbanised, issues relating to the ongoing integration policies and improving relationship between police and residents, as well as concerns about health and safety, loom especially large. There are undoubtedly many questions to be asked about favela, slum, dark and myriad other forms of modern tourist activity. Yet if exploitation and ethical unease is to be detected anywhere in Rio de Janeiro, it can find its mark just as easily on the beaches of Ipanema and Copacabana as it can in Rochina.

Right now, as Christopher Gaffney suggests,

> Rio de Janeiro has fully engaged the process of making itself into an Olympic City where the workers will stream down from the favelas to build sportive constellations that are intended for use by the international tourist class and the upper strata of Brazilian society.[43]

In other words, tourists don't need to travel to the favelas to form a parasitic and potentially voyeuristic part of the chronic inequalities of the city; they need only sit at the pavement cafes and in the restaurants, and in the glamorous bars of Leblon, where they will be waited on hand and foot by the poorest residents of the city.

Notes

1 Brasilia was listed by UNESCO in 1987, qualifying in two of the ten criteria (i and iv respectively). Regarded as an outstanding example of modernist urbanism, its blend of 'human creative genius' and significance 'in human history' is seen as having combined to produce a singular artistic achievement. See UNESCO, *The World's Heritage* (UNESCO: Paris, 2009), 280. Glenny would dispute my interpretation concerning Rio's later economic and cultural fortunes: 'Rio lost its status when Brasilia was made the country's capital in 1960. It has never quite recovered. Apart from the prestige, it also lost tens of thousands of government jobs to the brash newcomer. In the eighties and nineties, it also haemorrhaged entire industrial sectors to São Paulo': M. Glenny, *Nemesis: One Man and the Battle for Rio* (London: Bodley Head, 2015), 97.
2 Carvhalho evokes the feeling perfectly: 'When the Italian architect Lina Bo Bardi arrived in Rio de Janeiro in October 1946 she used a single word to describe her initial impression: *incanto* ('enchantment')': B. Carvalho, *Porous City: A Cultural History of Rio de Janeiro* (Liverpool: Liverpool University Press, 2013), ix.
3 The Portuguese Court, originally installed in Salvador da Bahia, found the city cramped and quickly relocated to Rio: 'When the court arrived in Rio de Janeiro, it elevated the colonial capital to the capital of the empire.' Increasingly entrenched, and 'possibly reluctant to trade the tropical splendour of Rio de Janeiro for Lisbon', they sat put. When João VI returned to Portugal in 1821 a 'little less than half of the Court and military that had accompanied the King to Brazil a decade earlier returned with him'. T. Meade, *A Brief History of Brazil* (New York: Checkmark, 2004), 74.
4 Changes across the favelas have been noted by several critics, with improved standards, greater facilities and higher living costs associated with Vidigal in particular. For detailed analysis see J. Cummings, "Confronting Favela Chic: The Gentrification of Informal Settlements in Rio de Janeiro, Brazil," in *Global Gentrifications: Uneven Development and Displacement*, ed. L. Lees, H. Bang Shin, and E. López-Morales (Bristol: Policy Press, 2015), 81–101.

5 For a very full analysis of the Pacification Schemes, and of the impact and recent controversies associated with Rochina favela in particular, see Glenny: 'A new long-term policy was needed to secure Rio's future, one that involved the state making amends for its disgraceful neglect of the favelas and Rio's poor over the previous several decades.' Glenny, *Nemesis*, 211.
6 For a full account of police and military brutality throughout the period of the Military Government, see J. Dassin, ed., *Torture in Brazil* (São Paulo: Vintage, 1986) [trans. J. Wright]. Prepared by the Archdiocese of São Paulo, this secretly prepared report graphically catalogues human rights abuses, including the abuse of pregnant women and children, the collusion of medics, and the perverse distortion of the law.
7 See L. Lumsdon and J. Swift, *Tourism in Latin America* (London: Continuum, 2001) for wide-ranging analysis of Rio's tourism potential. The recent interest in sustainable and eco-tourism offerings has been successfully exploited by a number of companies, with trips organised to the rainforests, the seasonal floodplains of the Pantanal, the Amazon, and others. However, some critics are more sceptical: 'environmental policy in Brazil is far from being considered a social or a developmental policy. It is weakly related – if related at all – to other public policies, and the environmental question is still largely understood from a traditional ecological and conservationist perspective.' V. del Rio, "Sustainability and Contemporary Urbanism in Brazil," *Focus* 6, no. 1 (2009): 40.
8 The beach, of course, as a physical interface, can invite wider interpretations; see B. Carvalho, "Mapping the Urbanized Beaches of Rio de Janeiro: Modernization, Modernity and Everyday Life," in *Journal of Latin American Cultural Studies*, vol. 16, no. 3 (2007): 329, who sees the beach as a site of potential radicalism and political activism, as a 'space of contestation' within the city, not just as a signifier of tourist frivolity.
9 J. Rochester, *The Candelária Massacre* (London: Vision, 2008), 4–5. For further discussion of police and military disregard for children, see G. Dimenstein, *Brazil: War on Children* (London: Latin American Bureau, 1991); S. Branford and B. Kucinski, *Brazil: Carnival of the Oppressed* (London: Latin American Bureau, 1995).
10 '[. . .] the police were not police in a conventional sense but a faction at war with other factions in the favelas': M. Glenny, *Nemesis*, 86.
11 For interesting discussion of drugs and criminality in Rio, and especially the 'divided city' thesis which ignores interconnections between favelas and state officials, see E. D. Arias, *Drugs and Democracy in Rio de Janeiro: Trafficking, Social Networks and Public Security* (Chapel Hill: University of North Carolina Press, 2006).
12 One might also add to the favela tour the fairly recent tourist interest in *Reveillon*, a celebration organised in honour of the Sea Goddess and held on Copacabana beach as part of the New Year festivities. Associated for a long time by middle-class Brazilians as a mainly Afro-Brazilian, and therefore lower-class, festival, it has grown considerably in the past thirty years, in Rio as well as at other coastal venues. See G. Greenfield, "Reveillon in Rio de Janeiro," *Event Management* 14, no. 4, 2010: 301–8.
13 'Ashkenazi Jews began arriving in Brazil in significant numbers during the late nineteenth century, fleeing czarist regimes': B. Carvalho, *Porous City*, 105.
14 Meade, *'Civilizing' Rio*, 34.
15 Carvalho, *Porous City*, 16.
16 Meade, *'Civilizing' Rio*, 36.
17 J. D. Needell, *A Tropical Belle Epoque: Elite Culture and Society in turn-of-the-century Rio de Janeiro* (Cambridge: CUP, 1987) 49.
18 Carvalho, *Porous City*, 28.
19 Ibid, 76.
20 Meade, *'Civilizing' Rio*, 43.
21 Ibid, 105.
22 Of course Brazil was not unique in this regard. For example, anti-vaccination resistance took place in England in the mid-nineteenth century, especially around the towns of

Ipswich, Henley and Mitford, while later vaccination controversies also flared up in the USA, where an Anti-Vaccination Society of America had been formed.
23 Meade, *'Civilizing' Rio*, 81.
24 'Rochina has the highest incidence of tuberculosis in all Rio state, and in some years the highest incidence in Brazil. At this time [early 2000s], some 55 Rochinhans contract the illness every month': Glenny, *Nemesis*, 19.
25 In this sense, favela tourists run absolutely contrary to the mainstream, looking for the very thing that 10 per cent of North American travellers cite as something to be avoided. For pre-Pacification scheme analysis of Brazilian identities and the security-related threat to tourism revenue, see A. Rezende-Parker, A. Morrison, and J. Ismail, "Dazed and Confused? An Exploratory Study of the Image of Brazil as a Travel Destination," *Journal of Vacation Marketing* 9, no. 3 (2003): 243–59; P. Tarlow and G. Santana, "Providing Safety for Tourists: A Study of a Selected Sample of Tourist Destinations in the United States and Brazil," *Journal of Travel Research* 40, no. 4 (2002): 424–31.
26 The present interest in urban tourism could not have been remotely understood, not even as recently as the 1980s, when Rio was mired in controversy: 'once praised for its natural beauties, today it stands for all the malaise related to drug crimes, marginality, poverty and pollution. A city that potential international tourists are advised to avoid – metaphorically seen as 'The Beast'; V. Del Rio, "Urban Design and Conflicting City Images of Brazil: Rio de Janeiro and Curitiba," in *Cities* 9, no. 4 (1992): 271.
27 Glenny, *Nemesis*, 96.
28 On terror and social unrest, see G. Velho, "The Challenge of Violence," in University of São Paulo, *Brazil: Dilemmas and Challenges* (São Paulo: University of São Paulo Press, 2001).
29 J. Rochester, *The Candelária Massacre*, 94. 'The events at Vigário Geral reiterated, were it needed, how far the state had lost control of the police – or possibly how far the state now sanctioned aberrant police activity.' M. Glenny, *Nemesis*, p. 85.
30 For discussion of the criticisms concerning government spending on the World Cup and Olympic Games, see M. Conde and T. Jazeel, "Kicking Off in Brazil: Manifesting Democracy," in *Journal of Latin American Studies* 22, no. 4 (2013): 437–50. Interestingly, the authors' analysis of the impact of the various 'manifestações' – public protests that began in São Paulo and quickly swept across the country – are not confined to the favelas and other poor suburbs, but have wider social acceptance.
31 M. Rolfes, "Slumming – empirical results and observational-theoretical considerations on the backgrounds of township, favela and slum tourism," in *Tourist Experience: Contemporary Perspectives*, ed. R. Sharpley and P. R. Stone (London: Routledge, 2011), 68–9.
32 See E. Mesker, "Community Policing of Rio's Favelas: State-Led Development or Market-Oriented Intervention," in *Brazil Emerging: Inequality and Emancipation*, ed. J. Nederveen Pieterse and A. Cardoso (London: Routledge, 2014), 138–53.
33 T. Caldeira, *City of Walls: Crime, Segregation, and Citizenship in São Paulo* (Berkeley: University of California Press, 2000), 78.
34 For discussion of the integration policies of the 1990s, and the *Favela-Bairro* programme in particular, see F. Luiz Lara, "Favela Upgrade in Brazil: A Reverse of Participatory Processes," in *Journal of Urban Design* 18, no. 4 (2013): 553–64.
35 Caldeira, *City of Walls*, 79.
36 Carvalho, *Porous City*, 13.
37 For generally sceptical views, see S. E. Novacich, "Minha Casa, Minha Vida Development," *The Rio Times*, May 17, 2011; R. Selvanayagam, "My House, My Life. Brazil Continues to Bring Very Little Positive Change," October 10, 2012, www.habitationfortheplanet.org/blog/2012/10/my-house-my-life-minha-casa-minha-vida-base-of-the-pyramid-housing-brazil/ (accessed 21 March, 2016).
38 See M. Karl, "The High Price of Rio de Janeiro's Olympic Glory," May 4, 2015, www.atlasnetwork.org/news/article/the-high-price-of-rio-de-janeiros-olympic-glory (accessed November 10, 2015).

39 P. Stone, "Dark Tourism and Significant Other Death: Towards a Model of Mortality Mediation," *Annals of Tourism Research*, vol. 39, no. 3 (2012): 1580.
40 A. Biran, Y. Poria, and G. Oren, "Sought Experiences at (Dark) Heritage Sites," *Annals of Tourism Research* 38, no. 3 (2011): 836.
41 C. Rojek, "Indexing, Dragging and the Social Construction of Tourist Sights," in *Touring Cultures: Transformations of Travel and Theory*, ed. C. Rojek and J. Urry (London: Routledge, 1997), 62.
42 Glenny, *Nemesis*, 74.
43 C. Gaffney, "Mega-events and Socio-spatial Dynamics in Rio de Janeiro, 1919–2016," *Journal of Latin American Geography* 9, no. 1 (2010): 28.

Select bibliography

AAP Bulletins. "Ceremonies Mark Lockerbie Anniversary." *Australian Associated Press Pty LT*, 2008.
Adams, P. *Geographies of Media and Communication*. Chichester: John Wiley, 2009.
Allen, M. J., and S. D. Brown. "Embodiment and Living Memorials: The Affective Labour of Remembering the 2005 London Bombings." *Memory Studies* 4, no. 3 (2011): 312–27.
Anderson, B. *Imagined Communities: Reflections on the Origin and Spread of Nationalism*. London: Verso, 1983.
Armitage, A. 2007, "Mutual Research Designs: Redefining Mixed Methods Research Design." Paper presented at the British Educational Research Association Annual Conference, Institute of Education, University of London, September 5–8 2007.
Ashworth, G. "Holocaust Tourism: the Experience of Krakow: Kazimierz." *International Research in Geographical and Environmental Education* 11, no. 4 (2002): 363–7.
Ashworth, G. "Heritage, Identity and Places: For Tourists and Host Communities." In *Tourism in Destination Communities*, edited by S. Singh, L. India, D. J. Timothy, R. K. Dowling, and E. Cowan, 79–99. London: Cabi, 2003.
Ashworth, G. J., and J. E. Tunbridge. "Old cities, new pasts: Heritage planning in selected cities of Central Europe." *Geojournal* 49 (1999): 105–16.
Ashworth, G. J., and B. Graham, eds. *Senses of Place: Senses of Time*. Aldershot: Ashgate, 2005.
Ashworth, G. J., and M. Kavaratzis, eds. *Towards Effective Place Brand Management: Branding European Cities and Regions*. Cheltenham: Edward Elgar, 2010.
Assmann, J. "Communicative and Cultural Memory." *New German Critique* 65, no. 2 (1995): 125–33.
Assmann, J. "Communicative and Cultural Memory." In *Cultural Memory Studies: An International and Interdisciplinary Handbook,* edited by A. Erll and A. Nunning, 109–19. Berlin: de Gruyter, 2008.
Atkinson, D. "Magical Corpses: Ballads, Intertextuality, and the Discovery of Murder." *Journal of Folklore Research* 36, no. 1 (1999): 1–29.
Babergh District Council. "Polstead Conservation Area Appraisal." Accessed June 2, 2015, www.babergh.gov.uk/assets/Uploads-BDC/Economy/Heritage/Con-Area-Apps/Polstead2012CAA.pdf
Baker, B. *Destination Branding for Small Cities: The Essentials for Successful Place Branding*. Oregon: Creative Leap Books, 2007.
Barrell, J. *The Dark Side of the Landscape: The Rural Poor in English Painting 1730–1840*. Cambridge: Cambridge University Press, 1980.

Select bibliography

Basu, P. *Narratives in a Landscape: Monuments and Memories of the Sutherland Clearances*. London: University College, 1997.

Basu, P. "Cairns in the Landscape." In *Landscapes beyond Land*, edited by A. Arnason, N. Ellison, J. Vergunst, and A. Whitehouse, 116–39. Oxford: Berghahn, 2012.

Beech, J. "The Enigma of Holocaust Sites as Tourism Attractions: The Case of Buchenwald." *Managing Leisure* 5, no. 1: (2000) 29–41.

Bell, Duncan S.A. "Mythscapes: Memory, Mythology, and National Identity." *British Journal of Sociology* 54, no. 1 (2003): 63–81.

Bender, Barbara. "Stonehenge – Contested Landscapes (Medieval to Present-day)." In *Landscape, Politics and Perspectives*, edited by B. Bender, 245–79. Oxford: Berg, 1993.

Bender, D. "Record 1.5 Million Visitors to Auschwitz in 2014, 70 Percent Under 18." The Algemeiner. Accessed 18 February 2015, www.algemeiner.com/2015/01/04/record-1-5-million-visitors-to-auschwitz-in-2014-70-percent-under-18-video/

Bennett, R., and S. Savani. "The Rebranding of City Places: An International Comparative Investigation." *International Public Management Review* 4, no. 2 (2003): 70–87.

Bermingham, A. *Landscape and Ideology: the English Rustic Tradition, 1740–1860*. London: Thames and Hudson, 1987.

Bertman, S. L. ed. *Grief and the Healing Arts: Creativity as Therapy*. Amityville, NY: Baywood, 1999.

Bingham, R. D., M. A. Heath, R. C. Lyon, M. Williams, J. Whicker, and C. K. Paine. "Memorializing Columbine." *Illness, Crisis and Loss* 17, no. 3 (2009): 223–41.

Bishop, P. *An Archetypal Constable: National Identity and the Geography of Nostalgia*. London: Athlone, 1995.

Black, G. *The Engaging Museum: Developing Museums for Visitor Involvement*. Abingdon: Psychology Press, 2005.

Bonanno, G. A., C. Rennicke, and S. Dekel, "Self-enhancement among High-exposure Survivors of the September 11th Terrorist Attack: Resilience or Social Maladjustment?" *Journal of Personality and Social Psychology* 88, no. 6 (2005): 984–98.

Branscombe, N. R., and B. Doosje, eds. *Collective Guilt*. Cambridge: Cambridge University Press, 2004.

Brants, C. "Guilty Landscapes: Collective Guilt and International Criminal Law." In *Cosmopolitan Justice and Its Discontents*, edited by C. Bailliet and K. Aas, 59–69. London: Routledge, 2011.

Bremer, John. 2012. *C.S. Lewis, Poetry, and the Great War 1914–1918*. Lanham, MD: Lexington Books.

Brennan, M. "Condolence Books: Language and Meaning in the Mourning for Hillsborough and Diana." *Death Studies* 32, no. 4 (2008): 326–51.

Briggs, A. *Victorian Things*. London: Penguin, 1990.

Britton, D. "Comfort in Cloth: The Syracuse University Remembrance Quilt." *Voices* 34, nos. 3/4 (2008): 3–8.

Britton, D. *Elegies of Darkness: Commemorations of the Bombing of Pan Am 103*. PhD diss., Syracuse University, 2008. ProQuest.

Brookes, G., J. A. Pooley, and J. Earnest. *Terrorism, Trauma and Psychology: A Multilevel Victim Perspective of the Bali Bombings*. Abingdon: Routledge, 2014.

Brown, Lorraine. "Tourism and Pilgrimage: Paying Homage to Literary Heroes." *International Journal of Tourism Research* 18, no. 2 (2016): 167–75.

Brubaker, J. R., G. R. Hayes, and P. Dourish, "Beyond the Grave: Facebook as a Site for the Expansion of Death and Mourning." *The Information Society* 29, no. 3 (2013): 152–63.

Bryman, A. *Social Research Methods*. Oxford: Oxford University Press, 2008.

Buber, R., J. Gadner, and L. Richards, eds. *Applying Qualitative Methods to Marketing Management Research*. London: Palgrave, 2004.

Bunting, Madeleine. *The Model Occupation*. London: BCA, 1995.

Butler, R., and Wantanee Suntikul, "Tourism and War: An Ill Wind?" In *Tourism and War*, edited by R. Butler and Wantanee Suntikul, 1–11. Abingdon: Routledge, 2013.

Carr, G. *Occupied Behind Barbed Wire*. Jersey: Jersey Heritage, 2010.

Carr, G. "Examining the Memorialscape of Occupation and Liberation: A Case Study from the Channel Islands." *International Journal of Heritage Studies* 18, no. 2 (2012): 174–93.

Carr, G., Paul Sanders, and Louise Willmot. *Protest, Defiance and Resistance in the Channel Islands: German Occupation 1940–1945*. London: Bloomsbury Academic, 2014.

Chung, C., C. Chung, and Y. M. Easthope. "Traumatic Stress and Death Anxiety among Community Residents Exposed to an Aircraft Crash." *Death Studies* 24, no. 8 (2000): 689–704.

Clifford, J. *Routes: Travel and Translation in the Late Twentieth Century*. Cambridge, MA: Harvard University Press, 1997.

Coats, A., and S. Ferguson. "Rubbernecking or Rejuvenation: Post Earthquake Perceptions and the Implications for Business Practice in a Dark Tourism Context." *Journal of Research for Consumers* 23 (2013): 32–64.

Cohen, D. "The Beautiful Female Murder Victim: Literary Genres and Courtship Practices in the Origins of a Cultural Motif, 1590–1850." *Journal of Social History* 31, no. 2 (1997): 277–306.

Cohen, F. *The Jews in the Channel Islands during German occupation, 1940–1945*. London: Jersey Heritage Trust/The Institute of Contemporary History and Wiener Library Ltd, 2000.

Collins, P. 2006. "The Molecatcher's Daughter." Accessed 26 May 2015, www.independent.co.uk/arts-entertainment/books/features/the-molecatchers-daughter-425522.html

Connerton, P. "Seven Types of Forgetting." *Memory Studies* 1, no. 1 (2008): 59–71.

Corr, C., and D. Corr, *Death and Dying, Life and Living*. Belmont, CA: Cengage Learning, 2012.

Cosgrove, D. *Social Formation and Symbolic Landscape*. Madison, WI: Wisconsin University Press, 1998.

Creswell, J. W. *Educational Research: Planning, Conducting, and Evaluating Quantitative and Qualitative Approaches to Research*. Upper Saddle River, NJ: Merrill/Pearson Education, 2002.

Creswell, J. W., and V. L. Plano Clark. *Designing and Conducting Mixed Methods Research*. London: Sage Publications, 2006.

Cruickshank, Charles. *The German Occupation of the Channel Islands*. London: Oxford University Press/Channel Islands: Guernsey Press, 2004 [1975].

Curtis, J. *An Authentic and Faithful History of the Mysterious Murder of Maria Marten*. London: Thomas Kelly, 1828.

Dachau Marketing Authority Representative. Interview with the author on 26 September 2011. [Mp3 recording in possession of author; for full transcript see Appendix 2].

Daniels, S. *Fields of Vision: Landscape Imagery and National Identity in England and the United States*. Oxford: Polity, 1993.

Dann, G. "Writing Out the Tourist in Space and Time." *Annals of Tourism Research* 26, no. 1 (1999): 159–87.

Dass-Brailsford, P. "Disasters." In *Social Death. The A–Z of Death and Dying: Social, Medical, and Cultural Aspects,* edited by M. Brennan, 161–66. ABC-CLIO, Santa Barbara, California, 2014.

Dauncey, H., and C. Tinker. "Media, Memory and Nostalgia in Contemporary France: Between Commemoration, Memorialisation, Reflection and Restoration." *Modern and Contemporary France* 23, no. 2 (2015): 135–45.
Dekel, I. "Ways of Looking: Observation and Transformation at the Holocaust Memorial, Berlin." *Memory Studies* 2, no. 1 (2009): 71–86.
Denscombe, M. *The Good Research Guide: For Small-scale Social Research Projects.* Milton Keynes: Open University Press, 2010.
Dickie, V. A. "Experiencing Therapy through Doing: Making Quilts." *OTJR: Occupation, Participation and Health* 31, no. 4 (2011): 209–15.
Dinnie, K. *City Branding: Theory and Cases.* London: Palgrave, 2011.
Distel, B., G. Hammermann, S. Zámecnik, J. Zarusky, and Z. Zofka, eds. *The Dachau Concentration Camp 1993–1945*, 4th ed. Munich: Edition Lipp Verlagsgesellschaft mbH, 2005.
Doka, K. J. "Memorialization, Ritual and Public Tragedy." In *Living with Grief: Coping with Public Tragedy*, edited by M. Lattanzi-Licht and K. J. Doka, 179–90. New York: Brunner-Routledge, 2003.
Doss, E. L. *The Emotional Life of Contemporary Public Memorials: Towards a Theory of Temporary Memorials*, vol. 3. Amsterdam: Amsterdam University Press, 2008.
Drvenkar, N., Banožić, M., and Živić, D. "Development of Memorial Tourism as a New Concept – Possibilities and Restrictions." *Tourism and Hospitality Management* 21, no. 1 (2015): 63–77.
Dunning, Richard. *The Friends of Lochnagar Newsletter*, 37, July 2000: n.p.
Dunton, J. W. R. "Quilt Making as a Socializing Measure." *American Journal of Physical Medicine and Rehabilitation* 16, no. 4 (1937): 275–8.
Edensor, T. "Performing Tourism, Staging Tourism: (Re)producing Tourist Space and Practice." *Tourist Studies* 1, no. 1 (2001): 59–81.
Elgood, G. "One Year On, Families Remember PanAm Disaster Victims." *Reuters News*, December 21, 1989.
Emerson, R. W. 1876. *English Traits.* Boston: James R. Osgood and Company.
Enander, A. "2. Human Needs and Behaviour in the Event of Emergencies and Social Crises." *Emergency Response Management in Today's Complex Society* 31 (2010): 31–73.
Eyre, A. "In Remembrance: Post-disaster Rituals and Symbols." *Australian Journal of Emergency Management* 14, no. 3 (Spring 1999): 23–9.
Eyre, A. "Remembering: Community commemoration after disaster." In *Handbook of Disaster Research*, edited by H. Rodriguez, E. L. Quarantelli, and R. Dynes, 441–55. New York: Springer, 2007.
Franklin, A. 2003. *Tourism: An Introduction.* London: Sage.
Fulbrook, M. 1999. *German National Identity after the Holocaust.* Cambridge: Polity Press.
G-2 Section U.S. Seventh Army (2008) *Dachau*, 5th ed. Bennington, VT: Merriam Press.
Geyer, M., and M. Latham. "The Place of the Second World War in German Memory and History." *New German Critique* 71 (1997): 5–40.
Ginns, M. *The Organisation Todt and the Fortress Engineers in the Channel Islands.* Jersey: Channel Islands Occupation Society Archive Book No.8, 2006.
Gough, P. "Contested Memories: Contested Site: Newfoundland and its Unique Heritage on the Western Front." *The Round Table* 96, no. 393 (2007): 693–705.
Gough, P. "Commemoration of War." In *The Ashgate Research Companion to Heritage and Identity*, edited by B. Graham and P. Howard, 215–29. Aldershot: Ashgate, 2008.
Gough, P. *A Terrible Beauty: British Artists in the First World War.* Bristol: Sansom and Co, 2010.

Graham, S. *The Challenge of the Dead: A Vision of the War and the Life of the Common Soldier in France, Seen Two Years afterwards between August and November.* London: Cassell, 1921.
Griffith, P. *Battle Tactics of the Western Front: The British Army's Art of Attack 1916–1918.* New Haven: Yale University Press, 1994.
Hamblen, J., and L. B. Slone. *What Are the Traumatic Stress Effects of Terrorism?* Virginia: National Center for Posttraumatic Stress Disorder, Department of Veteran Affairs, 2001.
Harding, S. G. *Feminism and Methodology: Social Science Issues.* London: John Wiley, 1988.
Harris, G. "Ten Years On, A Nation Remembers." *The Times,* 22 December 1998.
Harris, R. E. *Islanders Deported.* Ilford: CISS Publishing, 1979.
Hawdon, J. and Ryan, J. "Social Relations that Generate and Sustain Solidarity after a Mass Tragedy." *Social Forces* 89, no. 4 (2011): 1363–84.
Heeley, J. *Inside City Tourism: A European Perspective.* Bristol: Channel View Publications, 2011.
Hindley, C. *The History of the Catnach Press.* London: Hindley, 1869.
Historic England, 2015. List entry: Maria Marten's cottage. Accessed 2 June 2015. http://list.historicengland.org.uk/resultsingle.aspx?uid=1037040.
Hobsbawm, E., and G. Rude. *Captain Swing.* London: Verso, 2014.
Holt, T., and Valmai Holt. *Major and Mrs Holts Battlefield Guide to the Somme.* Barnsley: Pen and Sword, 1996.
Howard, P., I. Thompson, and E. Waterton. *The Routledge Companion to Landscape Studies.* London: Routledge, 2005.
Isaac, R. K., and L. Budryte-Ausiejiene. "Interpreting the Emotions of Visitors: A Study of Visitor Comment Books at the Grūtas Park Museum, Lithuania." In *Scandinavian Journal of Hospitality and Tourism,* ahead-of-print, 2015, pp. 1–25.
Johnson, M. *Ideas of Landscape.* Oxford: Blackwell, 2007.
Jones, D. "Thomas Campbell Foster and the Rural Labourer: Incendiarism in East Anglia in the 1840s." *Social History* 1, no. 1 (1976): 5–43.
Jordan, K. "What We Learned from the 9/11 First Anniversary." *The Family Journal* 11, no. 2 (2003): 110–16.
Kärki, F. U. "Norway's 2011 Terror Attacks: Alleviating National Trauma with a Large-scale Proactive Intervention Model." *Psychiatric Services* 66, no. 9 (2015): 910–12.
Kavaratzis, M. 2005. "Branding the City through Culture and Entertainment". Paper presented at the AESOP 2005 Conference, July 13–18, 2005, Vienna, Austria.
Kelman, I., and R. Dodds, "Developing a Code of Ethics for Disaster Tourism." *International Journal of Mass Emergencies and Disasters* 27, no. 3 (2009): 272–96.
Kempshall, C. 2012. "Forgetting the French." Accessed 21 March, 2016, http://ww1centenary.oucs.ox.ac.uk/?p=2296
Kenealy, D. "Commercial Interests and Calculated Compassion: The Diplomacy and Paradiplomacy of Releasing the Lockerbie Bomber." *Diplomacy and Statecraft* 23, no. 3, 2012 (55–75).
King, A. *Memorials of the Great War in Britain: The Symbolism and Politics of Remembrance.* London: Bloomsbury, 2014.
Knowles Smith, H. R. *The Changing Face of the Channel Islands Occupation: Record, Memory and Myth.* Basingstoke: Palgrave, 2007.
Kolb, B. M. *Tourism Marketing for Cities and Towns: Using Branding and Events to Attract Tourism.* Elsevier/Butterworth-Heinemann, 2006.

Koonz, C. "Between Memory and Oblivion: Concentration Camps in German Memory." In *Commemorations: The Politics of National Identity*, ed J. R. Gillis, 258–81. New Jersey: Princeton University Press, 1994.

Kropf, N. P., and B. L. Jones. "When Public Tragedies Happen: Community Practice Approaches in Grief, Loss, and Recovery." *Journal of Community Practice* 22, no. 3 (2014): 281–98.

Kübler-Ross, E., and D. Kessler. *On Grief and Grieving: Finding the Meaning of Grief through the Five Stages of Loss*. New York: Simon and Schuster, 2005.

KZ Gedenkstätte Dachau (2011) "Official Website." Accessed January 7, 2012, www.kz-gedenkstaette-dachau.de/

KZ Research Representative (2011) Interview with one of the authors on 26 September 2011. Dachau. [Mp3 recording in possession of authors].

Le Quesne, E. *The Occupation of Jersey Day by Day*. Jersey: La Haule, 1999.

Leaver, K. "Victorian Melodrama and the Performance of Poverty." *Victorian Literature and Culture* 27, no. 2 (1999): 443–56.

Leddy-Owen, C. 2014. "Reimagining Englishness: Race, Class, Progressive English Identities and Disrupted English Communities." *Sociology* 48, no. 6 (2014): 1123–38.

Lee, S. and D. Hoe-Lian Goh, "Gone too Soon": Did Twitter Grieve for Michael Jackson?" *Online Information Review* 37, no. 3 (2013): 462–78.

Legg, S. "Sites of Counter-memory: The Refusal to Forget and the Nationalist Struggle in Colonial Delhi." *Historical Geography* 33 (2005): 180–201.

Lennon, J. "Dark Tourism and Sites of Crime." In *Tourism and Crime*, edited by D. Botterill and T. Jones, 215–29. Oxford: Goodfellow, 2010.

Lennon, J., and M. Foley. *Dark Tourism: The Attraction of Death and Disaster*. London: Continuum, 2000.

Lennon, J., and M. Mitchell, "The Role of Sites of Death in Tourism." In *Remember Me: Constructing Immorality: Beliefs on Immorality, Life and Death*, edited by M. Mitchell, 167–79. Oxford: Routledge, 2007.

Lennon, J., and H. Smith, "Shades of Dark: Interpretation and Commemoration at the Sites of Concentration Camps at Terezin and Lety, Czech Republic." In *Representing the Unimaginable: Narratives of Disaster*, edited by A. Stock and C. Stott, 67–85. Oxford: Peter Lang, 2007.

Leslie, C. *Memoirs of the Life of John Constable*. London: Phaidon Press, 1951.

Lloyd, D. *Battlefield Tourism: Pilgrimage and the Commemoration of the Great War in Britain, Australia and Canada*. Oxford: Berg, 1998.

Lowenthal, D, "British National Identity and the English Landscape." *Rural History* 2, no. 2 (1991): 205–30.

Mabey, R. "Landscape: Terra Firma?" In *Towards a New Landscape*, edited by N. Alfrey, P. Barker, M. Drabble, R. Mabey, D. Matthews, and K. Raine, 62–9. London: Bernard Jacobson, 1993.

McCorristine, S. *William Corder and the Red Barn Murder*. London: Palgrave, 2014.

McElya, M. "Remembering 9/11's Pentagon Victims and Reframing History in Arlington National Cemetery." *Radical History Review* 111 (2011): 51–63.

McManus, R. *Death in a Global Age*. London: Palgrave, 2012.

Malik, O. "Aviation Security before and after Lockerbie." *Terrorism and Political Violence* 10, no. 3 (1998): 112–33.

Malone, C. "The Art of Remembrance: The Arts and Crafts Movement and the Commemoration of the British War Dead, 1916–1920." *Contemporary British History* 26, no. 1 (2012): 1–23.

Mannes, J. (2011) Visitor numbers of Dachau Painting Gallery, District Museum and New Gallery 2010 [email] (Personal communication, 11 October 2011).
Marcuse, H. *Legacies of Dachau: The Uses and Abuses of a Concentration Camp, 1933–2000*. Cambridge: Cambridge University Press, 2001.
Marcuse, H. "Reshaping Dachau for Visitors: 1933–2000." In *Horror and Human Tragedy Revisited: The Management of Sites and Atrocities for Tourism*, edited by G. Ashworth and R. Hartmann, 118–48. New York: Cognizant Communications, 2005.
Maskell, H. P. *The Soul of Picardy*. London: Ernest Benn, 1930.
Matar, K. I., and R. W. Thabit. *Lockerbie and Libya: A Study in International Relations*. North Carolina: McFarland, 2003.
Mazanec, J. A. ed. *International City Tourism*. London: Pinter, 1997.
Meisel, M. *Realizations: Narrative, Pictorial, and Theatrical Arts in Nineteenth-century England*. New Jersey: Princeton University Press, 2014.
Middlebrook, M. and M. Middlebrook, *The Somme Battlefields: A Comprehensive Guide from Crécy to the Two World Wars*. London: Penguin, 1991.
Miller, R. "Sales from the City." *Marketing* 11 (1997): 31–4.
Mitchell, D. "Cultural Landscapes: Just Landscapes or Landscapes of Justice?" *Progress in Human Geography* 27, no. 6 (2003): 787–96.
Mottram, R. H. *Through the Menin Gate*. London: Chatto and Windus, 1932.
Muskett, P. "The East Anglian Agrarian Riots of 1822." *Agricultural History Review* 32, no. 1 (1984): 1–13.
National Association of Areas of Outstanding Natural Beauty. "Dedham Vale." Accessed June 2, 2015, www.landscapesforlife.org.uk/dedham-vale-aonb.html
Neimeyer, R. A. *Techniques of Grief Therapy: Creative Practices for Counselling the Bereaved*. London: Routledge, 2012.
Nemeth, D. G., J. Kuriansky, K.P. Reeder, A. Lewis, K. Marceaux, T. Whittington, T. W. Olivier, N. E. May, and J. A. Safier. "Addressing Anniversary Reactions of Trauma through Group Process: the Hurricane Katrina Anniversary Wellness Workshops." *International Journal of Group Psychotherapy* 62, no. 1 (2012): 129–42.
Nicholls, S. "Disaster Memorials as Government Communication." *Australian Journal of Emergency Management* 21, no. 4 (2006): 36–43.
Nora, P. "Between Memory and History: Les *lieux de mémoire*." *Representations* 26 (1989): 7–24.
Olwig, K. "The Law of Landscape and the Landscape of Law: The Things that Matter." In *The Routledge Companion to Landscape Studies*, edited by P. Howard, I. Thompson, and E. Waterton, 253–62. London: Routledge, 2013.
Pantcheff, T. X. H. *Alderney: Fortress Island*. Chichester, Sussex: Phillimore, 1981.
Payne, C. *Toil and Plenty: Images of the Agricultural Landscape in England 1780–1890*. New Haven: Yale University Press, 1993.
Porter, Benjamin W., and Noel B. Salazar. "Heritage Tourism, Conflict, and the Public Interest: An Introduction." *International Journal of Heritage Tourism* 11, no. 5 (2005): 361–70.
Prior, R., and T. Wilson, "The First World War." *Journal of Contemporary History* 35 (2000): 319–28.
Reay, B. *The Last Rising of the Agricultural Labourers: Rural Life and Protest in Nineteenth-century England*. Oxford: Oxford University Press, 1990.
Reed, P. *Courcelette (Battleground Europe)*. Barnsley: Pen and Sword, 1998.
Reijnders, S. "Watching the Detectives: Inside the Guilty Landscapes of Inspector Morse, Baantjer and Wallander." *European Journal of Communication* 24, no. 2 (2009): 165–181.
Relph, E. *Place and Placelessness*. London: Pion, 1976.

Rensmann, L. "Collective Guilt, National Identity and Political Processes in Central Germany." In *Collective Guilt*, edited by N. R. Branscombe and B. Doosje, 169–91. Cambridge: Cambridge University Press, 2004.

Rigney, A. "Plenitude, Scarcity and the Circulation of Cultural Memory." *Journal of European Studies* 35, no. 1 (2005): 11–28.

Rigney, A. "The Dynamics of Remembrance: Texts between Monumentality and Morphing." In *Cultural Memory Studies: An International and Interdisciplinary Handbook*, edited by A. Erll and A. Nunning, 345–53. Berlin: de Gruyter, 2008.

Rodoski, K. H. "An Enduring Legacy of Hope: Twenty-five Years after the Pan Am 103 Terrorist Bombing, SU and Lockerbie, Scotland, Remain Committed to Honoring Those Lost and Moving Forward In Their Memory." *Syracuse University Magazine* 30, no. 3 (2013): 1–4.

Rogers, J. E. ed. *The Art of Grief: The Use of Expressive Arts in a Grief Support Group*. London: Routledge, 2011.

Rojek, C. *Ways of Escape*. Basingstoke: Macmillan, 1993.

Rojek, C. and Urry, J. *Touring Cultures: Transformations of Travel and Theory*. London: Routledge, 1997.

Rosenberg, J., and D. L. Peck. "Individual Reactions and Social Responses to Massive Loss of Life." In *Handbook of Death and Dying*, edited by Clifton D. Bryant, 223–36. Thousand Oaks, CA: SAGE, 2003.

Rosenthal, M. *Constable: The Painter and His Landscape*. New Haven, CT: Yale University Press, 1983.

Rosenthal, M. "This Green Unpleasant Land: Landscape and Contemporary Britain." In *The Place of Landscape: Concepts, Contexts, Studies*, edited by J. Malpas, 273–95. Cambridge, MA: MIT Press, 2011.

Salazar, N. B. "Building a Culture of Peace through Tourism: Reflexive and Analytical Notes and Queries." *Universitas Humanística* 62, no. 2 (2006): 319–36.

Samuel, Henry. "Western Front Battlefield Sees Most Detailed Ever Study". *The Telegraph*, June 11 2011. Accessed 21 March 2015, www.telegraph.co.uk/news/worldnews/europe/france/8569052/Western-Front-battlefield-sees-most-detailed-ever-study.html.

Samuel, R. "History Workshop, 1966–80." In *People's History and Socialist Theory*, edited by R. Samuel, 410–17. London: Routledge, 1981.

Sanders, P. *The British Channel Islands under German Occupation, 1940–1945*. Jersey: Société Jersiaise and Jersey Heritage Trust, 2005.

Sanders, P. "Narratives of Britishness: UK War Memory and Channel Islands Occupation Memory." In *Islands and Britishness: A global perspective*, edited by J. Matthews and D. Travers, 24–40. Newcastle Upon Tyne: Cambridge Scholars Publishing, 2012.

Santino, J. ed. *Spontaneous Shrines and the Public Memorialization of Death*. Basingstoke: Palgrave, 2006.

Saunders, N. J. "Material Culture and Conflict: The Great War, 1914–2003." In *Matters of Conflict: Material Culture, Memory and the First World War*, edited by N. J. Saunders, 5–26. London: Routledge, 2004.

Saunders, N. J. *Killing Time: Archaeology and the First World War*. Stroud: Sutton Publishing, 2007.

Seaton, A. V. "Guided by the Dark: From Thanatopsis to Thanatourism." *International Journal of Heritage Studies* 2 (1996): 234–44.

Seaton, Anthony V. "Another Weekend Away Looking for Dead Bodies. . .: Battlefield Tourism on the Somme and in Flanders." *Tourism Recreation Research* 25, no. 3 (2000): 63–77.

Seaton, A. "Purposeful Otherness: Approaches to the Management of Thanotourism." in *The Darker Side of Travel: The Theory and Practice of Dark Tourism*, edited by R. Sharpley and P. R. Stone, 75–109. Bristol: Channel View Publications, 2009.

Selman, P. and C. Swanwick. "On the Meaning of Natural Beauty in Landscape Legislation." *Landscape Research* 35, no. 1 (2010): 3–26.
Shields, R. "Political Tourism: Mapping Memory and the Future of Québec City." In *Mapping Tourism*, edited by S. P. Hanna and V. Del Casino, 1–27. Minneapolis: University of Minnesota Press, 2003.
Sierp, A. "Remembering to Forget? Memory and Democracy in Italy and Germany." Paper prepared for the XXIII Convegno SISP, September 17–19, 2009 Facoltà di Scienze Politische LUISS Guido Carli, Roma.
Simkins, P. "Introduction." In *The Somme: The Day-to-Day Account*, edited by C. McCarthy, 7–13. London: Armour Press, 1993.
Smith, Laurajane. *Uses of Heritage*. London: Routledge, 2006.
Stadt Dachau. "Imagefilm." Accessed 17 September 2011, www.youtube.com/user/StadtDachau
Stadt Dachau. "Official Website." Accessed 17 September 2011, www.dachau.de/
Stadt München. "Official Website." Accessed 20 January 2012, www.muenchen.de/
Stadt München. "Imagefilm." Accessed 20 January 2012, www.youtube.com/watch?v=Cuq55N94yiYandfeature=related
Steele, E. *Murder and Melodrama: The Red Barn Story on Stage*. College Park: University of Maryland, 2008.
Stevens, G. J., J. C. Dunsmore, K. E. Agho, M. R. Taylor, A. L. Jones, and B. Raphael, "Coping Support Factors among Australians Affected by Terrorism: 2002 Bali Bombing Survivors Speak." *The Medical Journal of Australia* 199, no. 11 (2013): 772–5.
Stillman, A. "Blogspot with 14–18 Project Manager, Andy Stillman." Accessed March 21, 2016, www.cwgc.org/media/78187/blogspot_with_andy_stillman.pdf
Stone, P. R. "A Dark Tourism Spectrum: Towards a Typology of Death and Macabre Related Tourist Sites, Attractions and Exhibitions." In *Tourism: An Interdisciplinary International Journal* 52, no. 2 (2006): 145–60.
Stone, P. R., and R. Sharpley. "Consuming Dark Tourism: A Thanatological Perspective." *Annals of Tourism Research* 35 (2008): 574–95.
Stour Valley Landscape Partnership. "Managing a Masterpiece." Accessed June 23, 2015, www.managingamasterpiece.org/
Sturdy Colls, C. "Holocaust Archaeology: Archaeological Approaches to Landscapes of Nazi Genocide and Persecution." *Journal of Conflict Archaeology* 7, no. 2 (2012): 71–105.
Taksa, L. "The Material Culture of an Industrial Artifact." *Historical Archaeology* 39, no. 3 (2005): 8–27.
Tarlow, P. "Dark Tourism – The Appealing 'Dark' Side of Tourism and More." In *Niche Tourism, Contemporary Issues, Trends and Cases*, edited by M. Novelli, 47–59. Oxford: Elsevier Butterworth-Heinemann, 2005.
Tashakkori, A. and C. Teddlie. *Mixed Methodology: Combining qualitative and quantitative approaches*. Thousand Oaks, CA: SAGE, 1998.
Thelwell, P. "Labour Camp Remains 'Should Become Occupation Memorial'." *Jersey Evening Post*, April 3, 2015: 1 and 8.
Uzzell, D. L. ed. *The Natural and Built Environment*. London: Belhaven, 1989.
Van Alphen, E. *Caught by History: Holocaust Effects in Contemporary Art, Literature, and Theory*. Stanford: Stanford University Press, 1997.
Van der Laarse, R. "Beyond Auschwitz? Europe's Terrorscapes in the Age of Postmemory." In *Memory and Postwar Memorials: Confronting the Violence of the Past*, edited by M. Silberman and F. Vatan, 71–92. New York: Palgrave, 2013.
Veenstra, J. "The New Historicism of Stephen Greenblatt." *History and Theory* 34, no. 3 (1995): 174–98.

Viejo-Rose, D. "Memorial Functions: Intent, Impact and the Right to Remember." *Memory Studies* 4, no. 4 (2011): 465–80.

Wallis, R. *Lockerbie: The Story and the Lessons*. Santa Barbara: Greenwood, 2001.

Walter, T. "Dark Tourism: Mediating between the Dead and the Living." In *The Darker Side of Travel: The Theory and Practice of Dark Tourism*, edited by R. Sharpley and P. R. Stone, 39–55. Bristol: Channel View, 2009.

Waterton, E., and S. Watson. *Semiotics of Heritage Tourism*. Bristol: Channel View, 2014.

Wight, C., and J. Lennon. "Towards an Understanding of Visitor Perceptions of Dark Attractions: The Case of the Imperial War Museum of the North, Manchester." *Journal of Hospitality and Tourism* 2, no. 2 (2004): 105–22.

Wilkinson, P. *The Lessons of Lockerbie: A Special Report on Aviation Security to Mark the First Anniversary of the Air Disaster* (No. 226). Research Institute for the Study of Conflict and Terrorism, 1989.

Williams, P. "Witnessing Genocide: Vigilance and Remembrance at Tuol Sleng and Choeng Ek." *Holocaust and Genocide Studies* 18, no. 2 (2004): 234–55.

Williams, R. *The Country and the City*. London: Hogarth Press, 1985.

Williams, R. *Culture and Materialism: Selected Essays*. London: Verso, 2005.

Wilson, Ross J. "The Popular Memory of the Western Front: Archaeology and European Heritage." In *Culture, Heritage and Representation: Perspectives on Visuality and the Past*, edited by E. Waterton and S. Watson, 75–90. Farnham: Ashgate, 2010.

Wilson, Ross J. *Landscapes of the Western Front: Materiality during the Great War*. London: Routledge, 2012.

Winter, J. M. *Sites of Memory, Sites of Mourning: The Great War in European Cultural History*, Cambridge: Cambridge University Press, 1995.

Winter, J. and E. Sivan, "Setting the Framework." In *War and Remembrance in the Twentieth Century*, edited by J. Winter and E. Sivan, 6–39. Cambridge: Cambridge University Press, 1999.

Wollaston, I. "Negotiating the Marketplace: The Role(s) of Holocaust Museums Today." *Journal of Modern Jewish Studies* 4, no. 1 (2005): 63–80.

Wombell, D. 2012. "Secret Underground War Site Revealed!" Accessed March 21, 2016, http://legerblog.co.uk/secret-underground-war-site-revealed/.

Wood, N. *Vectors of Memory: Legacies of Trauma in Postwar Europe*. Oxford and New York: Berg, 1999.

Woodthorpe, K. "Using Bereavement Theory to Understand Memorialising Behaviour." *Bereavement Care* 30, no. 2 (2011): 29–32.

Worsley, L. *A Very British Murder: The Story of a National Obsession*. London: Random House, 2013.

Wyschogrod, E. *An Ethics of Remembering: History, Heterology and the Nameless Others*. Chicago: University of Chicago Press, 1998.

Yrstad, V., and J. Schofield. "Remembering Høyblokka: The Government Building in Oslo, Norway – Confronting a Contemporary Heritage Dilemma." *The Historic Environment: Policy and Practice* 6, no. 1 (2015): 58–73.

Yudkina, A., and A. Sokolova, "Roadside Memorials in Contemporary Russia: Folk Origins and Global Trends." *Religion and Society in Central and Eastern Europe* 7, no. 1 (2014): 35–51.

Zamecnik, S. *Das war Dachau*, 2nd ed. Frankfurt am Main: Fischer Taschenbuch Verlag, 2010.

Zelizer, B. *Covering the Body: The Kennedy Assassination, the Media, and the Shaping of Collective Memory*. London: University of Chicago Press, 1992.

Index

academic history 91, 92
Alderney 100, 101, 102
amateur recordings 177, 180–3
amnesia *see* forgetting
Anne Frank House 75, 116–17
anniversaries 161, 162–3, 169
appropriateness: of attractions 27–8, 114; of locations 97; of memorials 160
archaeological excavations 101, 103–6
atrocity heritage 70, 81
Auschwitz-Birkenau 76–7, 78, 79, 112–13, 177
authenticity 29–30, 50, 77

Bermuda 19–21
black heritage 20
blogs 140–1, 150–1
body genre 176–7
branding 31–2
Brazil *see* Rio de Janeiro
Buchenwald 78–80
bunkers 96, 97–100
Burma railway 16
bus travel 148–52, 153–5, 156–7

Candelária Massacre 189
carceral tours 42–3
Channel Islands: heritage sites 19, 97–8; history 18, 96–7; legacy sites 98, 100–6; war narratives 17–19, 98–100
Charlie Hebdo 182–3
Chernobyl 5
Churchillian paradigm 99, 106
commemorative services 162–3
commercialisation 27–8, 114
commodification 7, 44–5, 110, 125
compassion 109, 115–16, 117
concentration/death camps: buildings 29–30, 76–7, 102; heritage contexts 26, 70–2; interpretation 29–30, 77–80; motivations 32–6, 73–6; tourism contexts 26–8, 72–3
Constable, John 84, 86–90, 92
contemporary memorials 161–4, 167
Copenhagen shooting 182–3
Corder, William 83–4
costumed interpretation 46–7
Crazy Guides Communism Tourism 179–80
creativity as therapy 166–7
cross-cultural interactions 198–9
cultural domain 59, 62
cultural landscapes 86, 89–90
cultural memory 86–7, 91–2
Curtis, James 83–4

Dachau: concentration camp 26–7, 28, 29; marketing 36–7; town 28, 30–1; visitor motivation 32–6
danger 148, 154, 157, 193
dark tourism: as concept 1–5, 12–14, 121–2, 126, 174–5; critique of 130–1, 135–7, 145; as perjorative term 135–6, 137, 145
darkness: continuum of 12, 13–14, 122–3; everyday 147–9, 157, 197–8; perceptions of 14–15, 96–8, 100, 106; tourism and 13–14, 137
death: attitudes towards 112; reflection on 131; ubiquitous 147, 149, 152
death camps *see* concentration/death camps
Dedham Vale 84, 87–8
demand-led dark tourism 23, 121, 123–6
Derby Gaol 48–50
desensitisation 41, 42, 76
dissonance 13, 14–16, 23, 70, 71–2
domain analysis 57–63

Dracula tourism: as dark tourism 121, 122, 125, 129–31; demand 125–9, 130–1; patrimony 63–6; supply 63, 123–5, 129–30
Dryfesdale *see* Lockerbie
duty to visit sites 112–13, 140–1, 143

education 30, 42–3, 48, 139, 168, 196
embodied experiences 148, 150, 200
emotion: media representations and 176–7; public outpourings 161; responses to sites 13, 110–11, 112, 141, 143–4
empathy 74, 75, 109, 113–14, 116; *see also* sympathy
entertainment 41–2, 44, 46–7, 50–1, 74, 128–9
ethics 3–5
everyday darkness 147–9, 157, 197–8
executions 49

fatality 56–7
favela tours 189–90, 193–201
fear 148, 150–1
fiction 62–4, 125–6, 177
forgetting: alternative narratives 18–19, 20–1, 75–6, 85–92, 97, 100; dark pasts 9, 70–1, 75–6, 177
fright tourism 127
funerary domain 59, 61–2

Galleries of Justice, Nottingham 45–8
genocide tourism 70, 134–5, 138–45; *see also* Holocaust tourism
geo-physical domain 57
geo-social domain 58
ghost tourism 43, 47–50
Gothic: patrimony 62–6; tourism 66, 126–8
governing myths 99
grief 160–1, 169
group identity 75
Guernsey 101–2
'guilty landscapes' 85–7, 91, 92–3

heritage: definition 97–8; dissonance 14–15; interpretation 77–80; tourism 14–17, 23; *see also* patrimony
'history from below' 91, 92
Holocaust tourism 26–8, 69–70, 72–5, 79–80, 135
'hot interpretation' 116–17

identity: construction 75, 119; dark/other 30, 37, 123; local communities 98, 103; visitors 72
imagination 126–8
impact of tourism 1–2, 16–17, 108, 110–11, 198–9
interpretation 27–8, 75–6, 77–80, 105, 116–17
Ireland 60–6
Italy 60–3

Jersey 17–19, 99, 100, 101–2, 103–6
Jewish genocide heritage 69, 71, 77–8
Jewish quarter, Krakow 15

Kanchanaburi, Thailand 16
Kazimierz, Krakow 15
Kigali Genocide Memorial (KGM) 134, 139, 140–5

labour camps 100–2
Lager Wick 103–6
landscape 84, 85–91
legacy sites 97–8, 100–6
lieux de mémoire/d'oubli 97, 100, 101; *see also* forgetting
literary tourism 63–4, 125–6
live environments 7–8, 197–8
liveness: at dark sites 178–80; definition 175–6; disseminated by tourists 180–2; of live events 182–3; through media 176–8
local community 16, 71–2, 76, 103
local history 165–7
Lockerbie: air disaster 160, 161–2; Dryfesdale Lodge Visitors' Centre 165–7; Garden of Remembrance 164–5

macabre, the 121–2, 127
Malta 21–2
management of sites 80–1
'Managing a masterpiece' project 89–90
marketing 30–2, 36–7, 47, 79
Marten, Maria 83–4
mea culpa tourism 72
media 41, 91, 176–7
memorials: learning and 29–30, 109–10; location of 96–7; for Lockerbie air disaster 161, 163–6; online 181; personal 164, 168; purpose of 134–5, 140–1, 143, 161, 170–1; selective 156–7

Index

mortality 56–7, 131, 200
motivation *see* producer motivation; visitor motivation
myths: folklore and 62, 64–5; governing 99

narratives: alternative 20, 85–92; dominant 87, 98–100; of favela tours 194–7; of genocide sites 143–4; and resolution 118
natural disasters 60–1, 74, 155–7
Nepal: civil war 152–5; earthquake 155–7; everyday tourism 148–52
Norway terror attacks 160–1, 174
Nottingham Galleries of Justice 45–8

online tourism sites 7–8, 153–4, 177, 181–2
Organisation Todt (OT) 100–1, 102–3
'otherness' 12, 42, 47, 63, 75
outsiders view 141

pacification scheme 196–7, 199
past-present-future links 116–17, 142, 144–5
patrimony: Gothic 62–6; thanatouristic 55–60
'peace through tourism' 110–11, 115–16, 118, 144
performance 44–5, 47–8, 91, 175–6, 178–80
photography 165–6
politico-legal domain 58, 61
politics and interpretation 77–8
Polstead, Suffolk 83–5, 92
popular culture 91
preservation 51, 74–5, 103
prison tourism: Derby Gaol 48–50; Nottingham Galleries of Justice 45–8; overview 40–3, 50–2; UK context 43–5
prisoners of war 20–1
producer motivation 74–6, 80–1
punishment 41–3, 49, 52

race 20–1
reconciliation 71, 135, 144
reconstruction of sites 29–30, 77
Red Barn murder 83–4, 91–3
reflection: on everyday life 109, 118–19, 144, 157; on human suffering 108–9, 113–14; opportunities for 114–15, 117, 200
religious domain 58, 62
remediations 176–7, 181

remembrance: healing and 161; interpretation and 92–3; to prevent future atrocities 75–6, 108–11, 144; selective 56, 71
Rio de Janeiro: favela tours 193–201; history 187–93; social inequality 190–1, 192–3
risk awareness 152, 154, 155
Rochina (favela) 188, 198, 199–200
Romania 123–6
Royal Naval Dockyard, Bermuda 20–1
Rwanda: genocide tourism 134–6; Kigali Genocide Memorial (KGM) 134, 138–45; memorials 139

sanitisation 77, 105–6, 115
screen tourism 126–7, 177
self, the 179–80, 200
sensation sites 27
Shevchenko, Inna 183
Sicily 60–3
slave labourers 100–1
slavery 20
social aims 75, 108, 110–11, 118–19, 184, 198–9
social inequality 20, 59, 157, 193–4, 198, 199
socio-economic domain 59
Stoker, Bram 63–6, 123
Stour Valley, Suffolk 83–4
structural amnesia 87; *see also* forgetting
suffering 115–16, 147
supernatural, the 47–50, 127–9
supply-led dark tourism 23, 72–3, 121, 123
sympathy 178–80; *see also* empathy
Syracuse University, New York State 162, 163, 169

thanatourism: definitions 27, 72–3, 121; as destination concept 66–7; patrimony 55–60
theatre 178–9
tradition 91–2
transitional places 96–7
Transylvania, Romania 63, 123–6
travel blogs 140–1, 150–1

Utøya, Norway 160–1, 174

vampires 63–6, 125, 127–9; *see also* Dracula tourism
victim tourism 72, 75
visitor experience 108, 109, 111–13, 140–2, 201

visitor motivation: discussion 6, 73–4, 123, 135–7, 174; research methods 32–4, 111–12, 140; research outcomes 34–6, 112–14, 130–1, 141, 142–3, 145
visitor responses 13–14, 116–18, 140–5, 167–8, 194
visual heritage 86, 88–9
voyeurism 112, 115, 135, 201

war zones 152–5
Weimar 79–80
Whitby, Yorkshire 63, 127–8
witnesses: live 178, 180, 183; media 175, 176, 180, 181; tourists as 40–1, 50, 180–1

YouTube 180–2